STUDY GUIDE

Karen C. Timberlake

Los Angeles Valley College

CHEMISTRY

An Introduction to General, Organic,
and Biological Chemistry

ELEVENTH EDITION

Prentice Hall

Boston   Columbus   Indianapolis   New York   San Francisco   Upper Saddle River

Amsterdam   Cape Town   Dubai   London   Madrid   Milan   Munich   Paris   Montréal   Toronto

Delhi   Mexico City   São Paulo   Sydney   Hong Kong   Seoul   Singapore   Taipei   Tokyo

Editor in Chief, Chemistry: Adam Jaworski
Marketing Manager: Erin Gardner
Associate Editor: Jessica Neumann
Editorial Assistant: Catherine Martinez
Marketing Assistant: Nicola Houston
Managing Editor, Chemistry and Geosciences: Gina M. Cheselka
Project Manager: Ed Thomas
Supplement Cover Designer: Paul Gourhan
Operations Specialist: Maura Zaldivar
Cover Photo Credit: Pink Flamingos sleeping. Corbis Photo

Printed in the United States of America

10   9   8   7   6   5   4   3   2   1

ISBN-13:   978-0-321-71942-3
ISBN-10:   0-321-71942-5

Prentice Hall
is an imprint of

www.pearsonhighered.com

Table of Contents

Preface

This Study Guide is intended to accompany *Chemistry*: *An Introduction to General, Organic, and Biological Chemistry,* Eleventh Edition. The purpose of this Study Guide is to provide you with additional learning resources that increase your understanding of the concepts. Each section in the Study Guide is correlated with a chapter in the text. Within each section, there are *Learning Goals*, *Guides to Problem Solving*, *Learning Exercises*, and *Answers* that promote an understanding of the chemical principles of the *Learning Exercise*. A *Checklist* of learning goals and a multiple-choice *Practice Test* provide a study of the entire chapter content. The complete Solutions to the odd-numbered problems in the text are now provided in a separate supplement, *Student Solutions Manual*.

I hope that this Study Guide will help you to learn chemistry. If you wish to make comments or corrections, or ask questions, you can send me an e-mail message at khemist@aol.com.

Karen C. Timberlake

To the Student

"One must learn by doing the thing;
though you think you know it, you
have no certainty until you try."

—Sophocles

Here you are in a chemistry class with your textbook in front of you. Perhaps you have already been assigned some reading or some problems to do in the book. Looking through the chapter, you may see words, terms, and pictures that are new to you. This may very well be your first experience with a science class like chemistry. At this point you may have some questions about what you can do to learn chemistry. This *Study Guide* is written with those considerations in mind.

Learning chemistry is similar to learning a new skill such as tennis or skiing or driving. If I asked you how you learn to play tennis or ski or drive a car, you would probably tell me that you would need to practice every day. It is the same with learning chemistry. Learning the chemical ideas and learning to solve the problems depends on the time and effort you invest in it. If you practice every day, you will find that learning chemistry is an exciting experience and a way to understand the current issues of the environment, health, and medicine.

Manage Your Study Time

I often recommend a study system in which you read one section of the text and immediately practice the questions and problems that go with it. In this way, you concentrate on a small amount of information and actively use what you learned to answer questions. This helps you to organize and review the information without being overwhelmed by the entire chapter. It is important to understand each section, because the concepts build like steps. Information presented in each chapter proceeds from the basic to the more complex. Perhaps you will only study three or four sections of the chapter. As long as you also practice doing some problems at the same time, the information will stay with you.

Form a Study Group

I highly recommend that you form a study group in the first week of your chemistry class. Working with your peers will help you use the language of chemistry. Schedule a time to meet each week to study and prepare to discuss problems. You will be able to teach some things to the other students in the group, and sometimes they will help you understand a topic that puzzles you. You won't always understand a concept right away. Your group will help you see your way through it. Most of all, a study group creates a strong support system whereby students can help each other complete the class successfully.

Go to Office Hours

Finally, go to your tutor's and/or professor's office hours. Your professor wants you to understand and enjoy learning this material and should have office hours. Often a tutor is assigned to a class or there are tutors available at your college. Don't be intimidated. Going to see a tutor or your professor is one of the best ways to clarify what you need to learn in chemistry.

Using this Study Guide

Now you are ready to sit down and study chemistry. Let's go over some methods that can help you learn chemistry. The features in this Study Guide were written specifically to help you understand and practice the chemical concepts that are presented in your class and in your text.

1. Study Goals

The *Study Goals* give you an overview of the chapter and what you can expect to accomplish when you complete your study and learning of a chapter.

2. Think About It

Each chapter has a group of *Think About It* questions that encourage you to think about some of the ideas and practical applications of the chemical concepts you are going to study. You may find that you already have knowledge of chemistry in some of the areas. That will be helpful to you. Other questions give you an overview of the chemistry ideas you will be learning.

3. Key Terms

Each chapter introduces *Key Terms*. As you complete the description of the *Key Terms*, you will have an overview of the topics you will be studying in that chapter.

4. Chapter Section Key Concepts

Each section of the chapter begins with a list of *Key Concepts* to illustrate the important ideas in that section. This summary of concepts is written to guide you through each of the learning activities. When you are ready to begin your study, read the matching section in the textbook and review the sample exercises in the text. There are also references to the *Tutorials*, S*elf Study Activities*, and *Case Studies* in the *MasteringChemistry* online learning resource.

5. Learning Exercises

The *Learning Exercises* give you an opportunity to practice problem solving related to the chemical principles in the chapter. Each set of *Learning Exercises* reviews a chemical principle. There is room for you to answer the questions or complete the exercise. The answers are found immediately following each *Learning Exercise*. Check your answers right away. If they don't match the answer in the Study Guide, go back to the textbook and review that section of the text again. It is important to make corrections before you go on. Chemistry involves a layering of skills such that each one is understood before the next one can be learned.

6. Checklist

Use the *Checklist* to review your understanding of the Study Goals. This gives you an overview of the major topics in the section. If something does not sound familiar, go back and review. One aspect of being a strong problem-solver is the ability to check your knowledge and understanding as you study.

7. Practice Test

A *Practice Test* is found at the end of each chapter. If the results of this test indicate that you know the material, you are ready to proceed to the next chapter. If, however, the results indicate further study is needed, you can repeat the Learning Exercises in the sections you still need to work on. Answers for all of the questions are included at the end of the Practice Test.

This book is dedicated to my husband Bill.
Thank you for your help, expertise, patience, and gourmet cooking
that make it possible for me to finish this book.

1
Chemistry and Measurements

Study Goals

- Define the term *chemistry* and identify substances as chemicals.
- Develop a study plan for learning chemistry.
- Write the names and abbreviations for the metric or SI units used in the measurement of length, volume, mass, temperature, and time.
- Convert a standard number to scientific notation.
- Distinguish between measured numbers and exact numbers.
- Determine the number of significant figures in a measurement.
- Round off a calculator answer to the correct number of significant figures.
- Write conversion factors from the quantities and units of equalities.
- Use metric and SI units, U.S. units, a percentage, ppm or ppb, and density as conversion factors.
- Convert the initial unit of a measurement to another unit.
- Calculate the density of a substance; use density to convert between mass and volume.

Think About It

1. Why can we say that the salt and sugar we use are chemicals?

2. What are some things you can do to help you study and learn chemistry?

3. What kind of device would you use to measure each of the following: your height, your weight, and the quantity of water to make soup?

4. When you make a measurement, why should you write down a number and a unit?

5. Why does oil float on water?

Key Terms

Match each of the following key terms with the correct definition.

    a. metric system   **b.** exact number   **c.** significant figures
    d. conversion factor   **e.** chemistry   **f.** density
    g. scientific notation

1. ___ all the numbers recorded in a measurement including the estimated digit

2. ___ a fraction that gives the quantities of an equality in the numerator and denominator

3. ___ a form of writing a number using a coefficient and a power of ten

4. ___ the study of substances and how they interact

5. ___ the relationship of the mass of an object to its volume, usually expressed in g/mL

6. ___ a number obtained by counting items or from a definition

7. ___ a decimal system of measurement used throughout the world

Answers   **1.** c   **2.** d   **3.** g   **4.** e   **5.** f   **6.** b   **7.** a

1.1 Chemistry and Chemicals

- Chemistry is the study of the composition, structure, properties, and reactions of matter.
- A chemical is a substance containing one type of material that has the same composition and properties.

♦ Learning Exercise 1.1

Indicate if each of the following is a chemical:

    a. _____ aluminum   **b.** _____ heat

    c. _____ sodium fluoride in toothpaste   **d.** _____ ammonium nitrate in fertilizer

    e. _____ time

Answers   **a.** yes   **b.** no   **c.** yes   **d.** yes   **e.** no

1.2 A Study Plan for Learning Chemistry

- Components of the text that promote learning include: *Looking Ahead, Learning Goals, Concept Checks, Sample Problems, Study Checks, Guides to Problem Solving (GPS), Chemistry Link to Health, Chemistry Link to the Environment, Chemistry Link to History, Chemistry Link to Industry, Questions and Problems, Concept Maps, Chapter Reviews, Key Terms, Understanding the Concepts, Additional Questions and Problems, Challenge Problems, Glossary/Index, Answers to Selected Questions and Problems,* and *Combining Ideas.*
- An active learner continually interacts with chemical concepts while reading the text and attending lecture.
- Working with a study group clarifies ideas and illustrates problem solving.

♦ **Learning Exercise 1.2**

Which of the following activities would be included in a successful study plan for learning chemistry?

    a. _____ attending lecture once in a while

    b. _____ working problems with friends from class

    c. _____ attending review sessions

    d. _____ planning a regular study time

    e. _____ not doing the assigned problems

    f. _____ going to the instructor's office hours

Answers   **a.** no   **b.** yes   **c.** yes   **d.** yes   **e.** no   **f.** yes

1.3 Units of Measurement

- In the sciences, physical quantities are described in units of the metric or International System (SI).
- In the metric system, length or distance is measured in meters (m), volume in liters (L), mass in grams (g), and temperature in Celsius degrees (°C) or kelvins (K).

♦ **Learning Exercise 1.3**

Match the measurement with the quantity in each of the following:

  1. length   **2.** mass   **3.** volume   **4.** temperature   **5.** time

    a. _____ 45 g   **b.** _____ 8.2 m   **c.** _____ 215 °C   **d.** _____ 45 L

    e. _____ 825 K   **f.** _____ 8.8 s   **g.** _____ 140 g   **h.** _____ 50 s

Answers   **a.** 2   **b.** 1   **c.** 4   **d.** 3   **e.** 4   **f.** 5
                 g. 2   **h.** 5

1.4 Scientific Notation

- A value written in scientific notation has three parts: a number that is at least 1 but less than 10 called a *coefficient*, a power of 10 and a unit of measurement.
- For numbers greater than 10, the decimal point is moved to the left to give a positive power of 10.
- For numbers less than 1, the decimal point is moved to the right to give a negative power of 10.

(MC)

Tutorial: Scientific Notation
Tutorial: Using Scientific Notation

Study Note

1. For a number greater than 10, the decimal point is moved to the left to give a number that is at least 1 but less than 10 and a positive power of 10. For a number less than 1, the decimal point is moved to the right to give a number that is at least 1 but less than 10 and a negative power of 10.

2. To express 2500. in scientific notation, move the decimal point 3 places to the left, so we write 2.500×10^3. The number 2.5×10^3 means that 2.5 is multiplied by 10^3 (1000).
$$2.5 \times 1000 = 2500$$

3. To express 0.082 in scientific notation, move the decimal point 2 places to the right, so we write 8.2×10^{-2}. The number 8.2×10^{-2} means that 8.2 is multiplied by 10^{-2} (0.01).
$$8.2 \times 0.01 = 0.082$$

♦ **Learning Exercise 1.4**

Write the following measurements in scientific notation:

a. 240 000 cm _____

b. 825 m _____

c. 2300 kg _____

d. 53 000 y _____

e. 0.002 m _____

f. 0.000 015 g _____

g. 0.084 kg _____

h. 0.000 24 s _____

Answers

a. 2.4×10^5 cm	b. 8.25×10^2 m	c. 2.3×10^3 kg	d. 5.3×10^4 y
e. 2×10^{-3} m	f. 1.5×10^{-5} g	g. 8.4×10^{-2} kg	h. 2.4×10^{-4} s

1.5 Measured Numbers and Significant Figures

- A measured number is obtained when you use a measuring device to determine a quantity.
- An exact number is obtained by counting items or from a definition that relates units in the same measuring system.
- There is uncertainty in every measured number, but not in exact numbers.
- Significant figures in a measured number are all the reported figures including the estimated digit.
- A number is a significant figure if it is

  a. not a zero
  b. a zero between nonzero digits
  c. a zero at the end of a decimal number

- Zeros written at the beginning of a number with a decimal point or zeros in a large number without a decimal point are not significant digits.

(MC)

Self Study Activity: Significant Figures
Tutorial: Counting Significant Figures

Study Note

Significant figures (abbreviated SFs) are all the numbers reported in a measurement including the estimated digit. Zeros are significant unless they are placeholders appearing at the beginning of a decimal number or at the end of a large number without a decimal point.

4.255 g (four SFs)   0.0042 m (two SFs)   46 500 L (three SFs)

♦ **Learning Exercise 1.5A**

Are the numbers in each of the following statements measured (M) or exact (E)?

a. _____ There are 7 days in 1 week.   **b.** _____ A concert lasts for 73 min.

c. _____ There are 1000 g in 1 kg.   **d.** _____ The potatoes have a mass of 2.5 kg.

e. _____ A student has 26 DVDs.   **f.** _____ The snake is 1.2 m long.

Answers   **a.** E (counted)   **b.** M (use a watch)   **c.** E (metric definition)
          d. M (use a balance)   **e.** E (counted)   **f.** M (use a ruler)

♦ **Learning Exercise 1.5B**

State the number of significant figures in the following measured numbers:

a. 35.24 g   _____   **b.** 0.000 080 m _____

c. 55 000 m   _____   **d.** 805 mL   _____

e. 5.025 L   _____   **f.** 0.006 kg   _____

g. 268 200 mm   _____   **h.** 25.0 °C   _____

Answers   **a.** 4   **b.** 2   **c.** 2   **d.** 3
           e. 4   **f.** 1   **g.** 4   **h.** 3

1.6 Significant Figures in Calculations

- In multiplication or division, the final answer must have the same number of significant figures as the measurement with the fewest significant figures.
- In addition or subtraction, the final answer must have the same number of decimal places as the measurement with the fewest decimal places.
- When evaluating a calculator answer, it is important to count the significant figures in the measurements, and round off the calculator answer properly.
- Answers in chemical calculations rarely use all the numbers that appear in the calculator. Exact numbers are not included in the determination of the number of significant figures (SFs).

Tutorial: Significant Figures in Calculations

Study Note

1. To round off a number, when the first digit to be dropped is 4 *or less*, keep the digits you need and drop all the digits that follow.

    Round off 42.8254 to 3 SFs → 42.8 (drop digits 254)

2. To round off a number, when the first digit dropped is 5 *or greater*, keep the proper number of digits and increase the last retained digit by 1.

    Round off 8.4882 to 2 SFs → 8.5 (drop digits 882, increase last retained digit by 1)

3. When rounding off large numbers without decimal points, maintain the value of the answer by adding nonsignificant zeros as placeholders.

    Round 356 835 to 3 SFs → 357 000 (drop digits 835, increase final digit by 1, add placeholder zeros)

♦ **Learning Exercise 1.6A**

Round off each of the following to give two significant figures:

  a. 88.75 m   _____   **b.** 0.002 923 g _____

  c. 50.525 g   _____   **d.** 1.6726 m   _____

  e. 0.001 058 kg   _____   **f.** 82.08 L   _____

Answers   **a.** 89 m   **b.** 0.0029 g   **c.** 51 g   **d.** 1.7 m
                 e. 0.0011 kg   **f.** 82 L

Study Note

1. An answer obtained from multiplying and dividing has the same number of significant figures as the measurement that has the smallest number of significant figures.
$$1.5 \quad \times \quad 32.546 = 48.819 \rightarrow 49 \;\; \textit{Answer rounded to two SFs}$$
Two SFs   *Five SFs*

2. An answer obtained from adding or subtracting has the same number of decimal places as the initial number with the fewest decimal places.
$$82.223 \quad + \quad 4.1 = 86.323 \rightarrow 86.3 \;\; \textit{Answer rounded to one decimal place}$$
Three decimal   *One decimal*
places   *place*

♦ **Learning Exercise 1.6B**

Solve each problem and give the answer with the correct number of significant figures:

  a. $1.3 \times 71.5 =$ **b.** $\dfrac{8.00}{4.00} =$

  c. $\dfrac{0.082 \times 25.4}{0.116 \times 3.4} =$ **d.** $\dfrac{3.05 \times 1.86}{118.5} =$

  e. $\dfrac{376}{0.0073} =$ **f.** $38.520 - 11.4 =$

  g. $4.2 + 8.15 =$ **h.** $102.56 + 8.325 - 0.8825 =$

Answers   **a.** 93   **b.** 2.00   **c.** 5.3   **d.** 0.0479
                 e. 52 000 (5.2×10^4)   **f.** 27.1   **g.** 12.4   **h.** 110.00

1.7 Prefixes and Equalities

- In the metric system, larger and smaller units use prefixes to change the size of the unit by factors of 10. For example, a prefix such as *centi* or *milli* preceding the unit meter gives a smaller length than a meter. A prefix such as *kilo* added to the unit of gram gives a unit that measures a mass that is 1000 times greater than a gram.
- An equality contains two units that measure the *same* length, volume, or mass.
- Some common metric equalities are: 1 m = 100 cm; 1 L = 1000 mL; 1 kg = 1000 g.
- Some useful metric–American equalities are: 2.54 cm = 1 inch; 1 kg = 2.20 lb; 946 mL = 1 qt
- Some of the most common metric (SI) prefixes are shown below:

Prefix	Symbol	Numerical Value	Scientific Notation	Equality
Prefixes That Increase the Size of the Unit				
tera	T	1 000 000 000 000	10^{12}	1 Tg $= 1 \times 10^{12}$ g 1 g $= 1 \times 10^{-12}$ Tg
giga	G	1 000 000 000	10^{9}	1 Gm $= 1 \times 10^{9}$ m 1 m $= 1 \times 10^{-9}$ Gm
mega	M	1 000 000	10^{6}	1 Mg $= 1 \times 10^{6}$ g 1 g $= 1 \times 10^{-6}$ Mg
kilo	k	1 000	10^{3}	1 km $= 1 \times 10^{3}$ m 1 m $= 1 \times 10^{-3}$ km
Prefixes That Decrease the Size of the Unit				
deci	d	0.1	10^{-1}	1 dL $= 1 \times 10^{-1}$ L 1 L $= 10$ dL
centi	c	0.01	10^{-2}	1 cm $= 1 \times 10^{-2}$ m 1 m $= 100$ cm
milli	m	0.001	10^{-3}	1 ms $= 1 \times 10^{-3}$ s 1 s $= 1 \times 10^{3}$ ms
micro	μ	0.000 001	10^{-6}	1 μg $= 1 \times 10^{-6}$ g 1 g $= 1 \times 10^{6}$ μg
nano	n	0.000 000 001	10^{-9}	1 nm $= 1 \times 10^{-9}$ m 1 m $= 1 \times 10^{9}$ nm
pico	p	0.000 000 000 001	10^{-12}	1 ps $= 1 \times 10^{-12}$ s 1 s $= 1 \times 10^{12}$ ps

Self Study Activity: The Metric System
Tutorial: SI Prefixes and Units

♦ **Learning Exercise 1.7A**

Match the items in column **A** with those from column **B**.

	A		B
1.	_____ megameter	**a.**	10^{-9} m
2.	_____ terameter	**b.**	0.1 m
3.	_____ decimeter	**c.**	10^{-6} m
4.	_____ millimeter	**d.**	10^{12} m
5.	_____ centimeter	**e.**	0.01 m
6.	_____ nanometer	**f.**	1000 m

7. _____ micrometer   g. 10^{-3} m

8. _____ kilometer   h. 10^6 m

Answers   1. h   2. d   3. b   4. g
               5. e   6. a   7. c   8. f

♦ **Learning Exercise 1.7B**

Place the units in each of the following in order from smallest to largest:

a. kilogram   milligram   gram

b. centimeter   kilometer   millimeter

c. dL   mL   L

d. kg   mg μg

Answers   a. milligram, gram, kilogram   b. millimeter, centimeter, kilometer
               c. mL, dL, L   d. μg, mg, kg

♦ **Learning Exercise 1.7C**

Complete the following metric relationships:

a. 1 L = _____ mL   b. 1 L = _____ dL

c. 1 m = _____ cm   d. 1 dL = _____ mL

e. 1 kg = _____ g   f. 1 cm = _____ mm

g. 1 mg = _____ μg   h. 1 dL = _____ L

i. 1 m = _____ mm   j. 1 cm = _____ m

Answers   a. 1000   b. 10   c. 100   d. 100   e. 1000
               f. 10   g. 1000   h. 0.1   i. 1000   j. 0.01

1.8 Writing Conversion Factors

- Conversion factors are used in a chemical calculation to change from one unit to another.
- Each conversion factor represents an equality expressed in the form of a fraction.
- Two forms of a conversion factor can be written for any equality. For example, the metric–U.S. equality 2.54 cm = 1 in. can be written as follows:

$$\frac{2.54 \text{ cm}}{1 \text{ in.}} \quad \text{and} \quad \frac{1 \text{ in.}}{2.54 \text{ cm}}$$

♦ **Learning Exercise 1.8A**

Write two conversion factors for each of the following pairs of units:

a. millimeters and meters   b. kilograms and grams

c. kilograms and pounds

d. seconds and minutes

e. centimeters and meters

f. milliliters and quarts

g. deciliters and liters

h. millimeters and centimeters

Answers

a. $\dfrac{1000 \text{ mm}}{1 \text{ m}}$ and $\dfrac{1 \text{ m}}{1000 \text{ mm}}$

b. $\dfrac{1000 \text{ g}}{1 \text{ kg}}$ and $\dfrac{1 \text{ kg}}{1000 \text{ g}}$

c. $\dfrac{2.20 \text{ lb}}{1 \text{ kg}}$ and $\dfrac{1 \text{ kg}}{2.20 \text{ lb}}$

d. $\dfrac{60 \text{ s}}{1 \text{ min}}$ and $\dfrac{1 \text{ min}}{60 \text{ s}}$

e. $\dfrac{100 \text{ cm}}{1 \text{ m}}$ and $\dfrac{1 \text{ m}}{100 \text{ cm}}$

f. $\dfrac{946 \text{ mL}}{1 \text{ qt}}$ and $\dfrac{1 \text{ qt}}{946 \text{ mL}}$

g. $\dfrac{10 \text{ dL}}{1 \text{ L}}$ and $\dfrac{1 \text{ L}}{10 \text{ dL}}$

h. $\dfrac{10 \text{ mm}}{1 \text{ cm}}$ and $\dfrac{1 \text{ cm}}{10 \text{ mm}}$

Study Note

1. Sometimes, a statement within a problem gives an equality that is only true for that problem. Then conversion factors can be written that are true only for that problem. For example, a problem states that there are 50 mg of vitamin B in a tablet. The conversion factors are written as

$$\frac{1 \text{ tablet}}{50 \text{ mg vitamin B}} \quad \text{and} \quad \frac{50 \text{ mg vitamin B}}{1 \text{ tablet}}$$

2. If a problem gives a percentage (%), it can be stated as parts per 100 parts. For example, a candy bar contains 45% by mass chocolate. This percentage (%) equality can be written as conversion factors using the same mass unit such as grams.

$$\frac{45 \text{ g chocolate}}{100 \text{ g candy bar}} \quad \text{and} \quad \frac{100 \text{ g candy bar}}{45 \text{ g chocolate}}$$

3. When a problem gives ppm or ppb, it can be stated as parts per million (ppm), which is mg/kg, or parts per billion (ppb), which is μg/kg. For example, the level of nitrate in the Los Angeles water supply is 2.5 ppm. This ppm can be written as conversion factors using mg/kg.

$$\frac{2.5 \text{ mg of nitrate}}{1 \text{ kg of water}} \quad \text{and} \quad \frac{1 \text{ kg of water}}{2.5 \text{ mg of nitrate}}$$

Learning Exercise 1.8B

Write two conversion factors for each of the following statements:

    a. A cheese contains 55% fat by mass.   **b.** In the city, a car gets 14 mi/gal.

    c. A 125-g steak contains 45 g of protein.   **d.** 18-karat pink gold contains 25% copper by mass.

    e. A cadmium level of 1.8 ppm in food causes liver damage in rats.   **f.** A water sample contains 5.4 ppb of arsenic.

Answers

    a. $\dfrac{55 \text{ g fat}}{100 \text{ g cheese}}$ and $\dfrac{100 \text{ g cheese}}{55 \text{ g fat}}$ **b.** $\dfrac{14 \text{ mi}}{1 \text{ gal}}$ and $\dfrac{1 \text{ gal}}{14 \text{ mi}}$

    c. $\dfrac{45 \text{ g protein}}{125 \text{ g steak}}$ and $\dfrac{125 \text{ g steak}}{45 \text{ g protein}}$ **d.** $\dfrac{25 \text{ g copper}}{100 \text{ g pink gold}}$ and $\dfrac{100 \text{ g pink gold}}{25 \text{ g copper}}$

    e. $\dfrac{1.8 \text{ mg cadmium}}{1 \text{ kg food}}$ and $\dfrac{1 \text{ kg food}}{1.8 \text{ mg cadmium}}$ **f.** $\dfrac{5.4 \text{ } \mu\text{g arsenic}}{1 \text{ kg water}}$ and $\dfrac{1 \text{ kg water}}{5.4 \text{ } \mu\text{g arsenic}}$

1.9 Problem Solving

- Conversion factors from metric equalities, U.S. relationships, percent values, and density can be used to change a quantity expressed in one unit to a quantity expressed in another unit.
- The process of solving a problem with units requires the conversion of the initial unit to one or more units until the final unit needed for the answer is obtained.

Guide to Problem Solving Using Conversion Factors	
Step 1	State the given and needed quantities.
Step 2	Write a plan to convert the given unit to the needed unit.
Step 3	State the equalities and conversion factors needed to cancel units.
Step 4	Set up problem to cancel units and calculate answer.

Example: How many liters is 2850 mL?

        Step 1: State the given and needed quantities.

            Given 2850 mL   **Need** liters

        Step 2: Write a plan to convert the given unit to the needed unit.

            Milliliters $\xrightarrow{\text{Metric factor}}$ liters

Step 3: State the equalities and conversion factors needed to cancel units.

$$1 \text{ L} = 1000 \text{ mL}$$

$$\frac{1 \text{ L}}{1000 \text{ mL}} \quad \text{and} \quad \frac{1000 \text{ mL}}{1 \text{ L}}$$

Step 4: Set up problem to cancel units and calculate answer.

$$2850 \ \cancel{\text{mL}} \times \frac{1 \text{ L}}{1000 \ \cancel{\text{mL}}} = 2.85 \text{ L}$$

(MC)

Tutorial: Metric Conversions

♦ **Learning Exercise 1.9A**

Use metric conversion factors to solve the following problems:

a. 189 mL = _____ L   b. 2.7 cm = _____ mm

c. 0.0025 L = _____ mL   d. 76 mg = _____ g

e. How many meters tall is a person whose height is 175 cm?

f. There are 285 mL in a cup of tea. How many liters is that?

g. If a ring contains 13 500 mg of gold, how many grams of gold are in the ring?

h. You walked 1.5 km on the treadmill at the gym. How many meters did you walk?

Answers	a. 0.189 L	b. 27 mm	c. 2.5 mL	d. 0.076 g
	e. 1.75 m	f. 0.285 L	g. 13.5 g	h. 1500 m

♦ **Learning Exercise 1.9B**

Use metric–U.S. conversion factors to solve the following problems:

a. 18 in. = _____ cm   b. 4.0 qt = _____ L

c. 275 mL = _____ qt   d. 1300 mg = _____ lb

e. 150 lb = _____ kg   f. 840 g = _____ lb

g. 15 ft = _____ cm   h. 8.50 oz = _____ g

Answers	a. 46 cm	b. 3.8 L	c. 0.291 qt	d. 0.0029 lb
	e. 68 kg	f. 1.9 lb	g. 460 cm	h. 241 g

Study Note

1. For setups that require a series of conversion factors, it is helpful to write a plan first. Work from the given unit to the needed unit. Then use a conversion factor for each unit change.

$$\text{Given unit} \longrightarrow \text{unit (1)} \longrightarrow \text{unit (2) needed unit}$$

2. To convert from one unit to another, select conversion factors that cancel the given unit and provide the needed unit for the problem.

$$\cancel{\text{Given unit}} \times \frac{\text{unit (1)}}{\cancel{\text{given unit}}} \times \frac{\text{unit (2)}}{\cancel{\text{unit (1)}}} \quad \text{needed unit (2)}$$

Tutorial: Unit Conversions

Tutorial: Using Percentage as a Conversion Factor

Tutorial: Introduction to Unit Analysis Method

♦ Learning Exercise 1.9C

Use conversion factors to solve the following problems:

a. A piece of plastic tubing measures 120 mm. What is the length of the tubing in inches?

b. A statue weighs 245 pounds. What is the mass of the statue in kilograms?

c. Your friend's height is 6 ft 3 in. What is your friend's height in meters?

d. In a triple-bypass surgery, a patient requires 3.00 pints of whole blood. How many milliliters of blood were given?

e. A doctor orders 0.450 g of a sulfa drug. On hand are 150-mg tablets. How many tablets are needed?

f. A hand sanitizer contains 62% ethanol by volume. How many milliliters of ethanol are in a 1.18 L-bottle of hand sanitizer?

Answers	**a.** 4.7 in.	**b.** 111 kg	**c.** 1.9 m
	d. 1420 mL	**e.** 3 tablets	**f.** 730 mL

1.10 Density

• The density of a substance is a ratio of its mass to its volume, usually in units of g/mL or g/cm³. For example, the density of sugar is 1.59 g/mL and silver is 10.5 g/mL.

$$\text{Density} = \frac{\text{mass of substance}}{\text{volume of substance}}$$

• The volume of 1 mL is equal to the volume of 1 cm³.

<table>
<tr><td colspan="2" align="center">**Guide to Using Density**</td></tr>
<tr><td>**Step 1**</td><td>State the given and needed quantities.</td></tr>
<tr><td>**Step 2**</td><td>Write a plan to calculate the needed quantity.</td></tr>
<tr><td>**Step 3**</td><td>Write equalities and their conversion factors including density.</td></tr>
<tr><td>**Step 4**</td><td>Set up problem to calculate needed quantity.</td></tr>
</table>

(MC)

Tutorial: Density and Specific Gravity

<table>
<tr><td colspan="2" align="center">**Guide to Calculating Density**</td></tr>
<tr><td>**Step 1**</td><td>State the given and needed quantities.</td></tr>
<tr><td>**Step 2**</td><td>Write the density expression.</td></tr>
<tr><td>**Step 3**</td><td>Express mass in grams and volume in milliliters (mL) or cm^3.</td></tr>
<tr><td>**Step 4**</td><td>Substitute mass and volume into the density expression and calculate the density.</td></tr>
</table>

<table>
<tr><td align="center">**Study Note**</td></tr>
</table>

Density can be used as a conversion factor between the mass (g) and volume (mL) of a substance. The density of silver is 10.5 g/mL. What is the mass of 6.0 mL of silver?

$$6.0 \text{ mL silver} \times \frac{10.5 \text{ g silver}}{1 \text{ mL silver}} = 63 \text{ g of silver}$$

Density factor

What is the volume of 25 g of olive oil (D = 0.92 g/mL)?

$$25 \text{ g olive oil} \times \frac{1 \text{ mL olive oil}}{0.92 \text{ g olive oil}} = 27 \text{ mL of olive oil}$$

Density factor inverted

♦ **Learning Exercise 1.10**

Calculate the density or use density as a conversion factor to solve each of the following:

a. What is the density (g/mL) of glycerol if a 200.-mL sample has a mass of 252 g?

b. A person with diabetes may produce 5 to 12 L of urine per day. Calculate the density of a 100.0 mL-urine sample that has a mass of 100.2 g.

c. A small solid has a mass of 5.5 oz. When placed in a graduated cylinder with a water level of 25.2 mL, the object causes the water level to rise to 43.8 mL. What is the density of the object in g/mL?

d. A sugar solution has a density of 1.20 g/mL. What is the mass, in grams, of 0.250 L of the solution?

e. A piece of pure gold weighs 0.26 lb. If gold has a density of 19.3 g/mL, what is the volume, in mL, of the piece of gold?

f. Diamond has a density of 3.52 g/mL. What is the specific gravity of diamond?

g. A salt solution has a density of 1.15 g/mL and a volume of 425 mL. What is the mass, in grams, of the solution?

h. A 50.0-g sample of a glucose solution has a density of 1.28 g/mL. What is the volume, in liters, of the sample?

Answers	**a.** 1.26 g/mL	**b.** 1.002	**c.** 8.4 g/mL	**d.** 300. g
	e. 6.1 mL	**f.** 3.52	**g.** 489 g	**h.** 0.0391 L

Checklist for Chapter 1

You are ready to take the Practice Test for Chapter 1. Be sure that you have accomplished the following learning goals for this chapter. If you are not sure, review the section listed at the end of the goal. Then apply your new skills and understanding to the Practice Test.

After studying Chapter 1, I can successfully:

_____ Describe a substance as a chemical (1.1).

_____ Design a study plan for successfully learning chemistry (1.2).

_____ Write the names and abbreviations for the metric (SI) units of measurement (1.3).

_____ Write large or small numbers using scientific notation (1.4).

_____ Identify a number as a measured number or an exact number (1.5).

_____ Count the number of significant figures in measured numbers (1.6).

_____ Report an answer with the correct number of significant figures (1.6).

_____ Write a metric equality from the numerical values of metric prefixes (1.7).

_____ Write two conversion factors for an equality (1.8).

_____ Use a conversion factor to change from one unit to another (1.9).

_____ Calculate the density of a substance; use the density to calculate the mass or volume (1.10).

Practice Test for Chapter 1

Instructions: Select the letter preceding the word or phrase that best answers the question.

1. Which of the following would be described as a chemical?
   A. sleeping   **B.** salt   **C.** singing
   D. listening to a concert   **E.** energy

2. Which of the following is not a chemical?
   A. wool   **B.** sugar   **C.** feeling cold
   D. salt   **E.** vanilla

For questions 3 through 7, answer yes or no.

To learn chemistry successfully, I will:

3. ____ work the problems in the chapter and check answers

4. ____ attend some lectures, but not all

5. ____ form a study group

6. ____ set up a regular study time

7. ____ wait until the night before the exam to start studying

8. Which of the following is a metric measurement of volume?
   A. kilogram   **B.** kilowatt   **C.** kiloliter   **D.** kilometer   **E.** kiloquart

9. The measurement 24 000 g written in scientific notation is
   A. 24 g   **B.** 24×10^3 g   **C.** 2.4×10^3 g   **D.** 2.4×10^{-3} g   **E.** 2.4×10^4 g

10. The measurement 0.005 m written in scientific notation is
   A. 5 m   **B.** 5×10^{-3} m   **C.** 5×10^{-2} m   **D.** 0.5×10^{-4} m   **E.** 5×10^3 m

11. The measured number in the following is
   A. 1 book   **B.** 2 cars   **C.** 4 flowers   **D.** 5 rings   **E.** 45 g

12. The number of significant figures in 105.4 m is
   A. 1   **B.** 2   **C.** 3   **D.** 4   **E.** 5

13. The number of significant figures in 0.000 82 g is
   A. 1   **B.** 2   **C.** 3   **D.** 4   **E.** 5

14. The calculator answer 5.78052 rounded to two significant figures is
   A. 5   **B.** 5.7   **C.** 5.8   **D.** 5.78   **E.** 6.0

15. A calculator answer of 3486.512 when rounded to three significant figures is
   A. 4000   **B.** 3500   **C.** 349   **D.** 3487   **E.** 3490

16. The correct answer for the problem $16.0 \div 8.0$ is
   A. 2   **B.** 2.0   **C.** 2.00   **D.** 0.2   **E.** 5.0

17. The correct answer for the problem $58.5 + 9.158$ is
   A. 67   **B.** 67.6   **C.** 67.7   **D.** 67.66   **E.** 67.658

18. The correct answer for the problem $\dfrac{2.5 \times 3.12}{4.6}$ is
   A. 0.54   **B.** 7.8   **C.** 0.85   **D.** 1.7   **E.** 1.69

19. Which of these prefixes has the largest value?
   A. *centi*   **B.** *deci*   **C.** *milli*   **D.** *kilo*   **E.** *micro*

20. What is the decimal equivalent of the prefix *centi?*
    A. 0.001   **B.** 0.01   **C.** 0.1   **D.** 10   **E.** 100

21. Which of the following is the smallest unit of measurement?
    A. gram   **B.** milligram   **C.** kilogram   **D.** decigram   **E.** centigram

22. Which volume is the largest?
    A. mL   **B.** dL   **C.** cm^3 **D.** L   **E.** kL

23. Which of the following is a conversion factor?
    A. 12 in.   **B.** 3 ft   **C.** 20 m   **D.** $\dfrac{1000 \text{ g}}{1 \text{ kg}}$ **E.** 2 cm^3

24. Which is the correct conversion factor that relates milliliters and liters?
    A. $\dfrac{1000 \text{ mL}}{1 \text{ L}}$ **B.** $\dfrac{100 \text{ mL}}{1 \text{ L}}$ **C.** $\dfrac{10 \text{ mL}}{1 \text{ L}}$ **D.** $\dfrac{0.01 \text{ mL}}{1 \text{ L}}$ **E.** $\dfrac{0.001 \text{ mL}}{1 \text{ L}}$

25. Which is the correct conversion factor for the relationship between millimeters and centimeters?
    A. $\dfrac{1 \text{ mm}}{1 \text{ cm}}$ **B.** $\dfrac{10 \text{ mm}}{1 \text{ cm}}$ **C.** $\dfrac{100 \text{ cm}}{1 \text{ mm}}$ **D.** $\dfrac{100 \text{ mm}}{1 \text{ cm}}$ **E.** $\dfrac{10 \text{ cm}}{1 \text{ mm}}$

26. 294 mm is equal to
    A. 2940 m   **B.** 29.4 m   **C.** 2.94 m   **D.** 0.294 m   **E.** 0.0294 m

27. The handle on a tennis racket measures 4.5 in. What is its size in centimeters?
    A. 11 cm   **B.** 1.8 cm   **C.** 0.56 cm   **D.** 450 cm   **E.** 15 cm

28. What is the volume of 65 mL in liters?
    A. 650 L   **B.** 65 L   **C.** 6.5 L   **D.** 0.65 L   **E.** 0.065 L

29. What is the mass in kg of a 22-lb turkey?
    A. 10. kg   **B.** 48 kg   **C.** 10 000 kg   **D.** 0.048 kg   **E.** 22 000 kg

30. The number of milliliters in 2 deciliters is
    A. 20 mL   **B.** 200 mL   **C.** 2000 mL   **D.** 20 000 mL   **E.** 500 000 mL

31. A person who is 5 ft 4 in. tall would be
    A. 64 m   **B.** 25 m   **C.** 14 m   **D.** 1.6 m   **E.** 1.3 m

32. How many oz are in 1500 g? (1 lb = 16 oz)
    A. 94 oz   **B.** 53 oz   **C.** 24 000 oz   **D.** 33 oz   **E.** 3.3 oz

33. How many qt of orange juice are in 255 mL of juice?
    A. 0.255 qt   **B.** 270 qt   **C.** 236 qt   **D.** 0.270 qt   **E.** 0.400 qt

34. An order for a patient calls for 0.020 g of medication. On hand are 4-mg tablets. How many tablets are needed for the patient?
    A. 2 tablets   **B.** 4 tablets   **C.** 5 tablets   **D.** 8 tablets   **E.** 200 tablets

35. A doctor orders 1500 mg of a sulfa drug. The tablets in stock are 0.500 g. How many tablets are needed?
    A. 1 tablet   **B.** 1½ tablets   **C.** ⅓ tablet   **D.** 2½ tablets   **E.** 3 tablets

36. What is the density of a bone with a mass of 192 g and a volume of 120 cm^3?
    A. 0.63 g/mL   **B.** 1.4 g/cm^3 **C.** 1.6 g/cm^3 **D.** 1.9 g/cm^3 **E.** 2.8 g/cm^3

37. How many milliliters of a salt solution with a density of 1.8 g/mL are needed to provide 400. g of salt solution?
    A. 220 mL   **B.** 22 mL   **C.** 720 mL   **D.** 400 mL   **E.** 4.5 mL

38. The density of a solution is 0.85 g/mL. Its specific gravity is
    A. 222 mL   **B.** 8.5   **C.** 0.85 mL   **D.** 1.2   **E.** 0.85

39. Three liquids have densities of 1.15 g/mL, 0.79 g/mL and 0.95 g/mL. When the liquids, which do not mix, are poured into a graduated cylinder, the liquid at the top is the one with a density of
    A. 1.15 g/mL   **B.** 1.00 g/mL   **C.** 0.95 g/mL   **D.** 0.79 g/mL   **E.** 0.16 g/mL

40. A sample of oil has a mass of 65 g and a volume of 80.0 mL. What is its density?
    A. 1.5 g/mL   **B.** 1.4 g/mL   **C.** 1.2 g/mL   **D.** 0.90 g/mL   **E.** 0.81 g/mL

41. What is the mass of a 10.0 mL sample of urine with a density of 1.04 g/mL?
    A. 104 g   **B.** 10.4 g   **C.** 1.04 g   **D.** 1.40 g   **E.** 9.62 g

42. Ethyl alcohol has a density of 0.790 g/mL. What is the mass, in grams, of 0.250 L of the alcohol?
    A. 198 g   **B.** 158 g   **C.** 3.95 g   **D.** 0.253 g   **E.** 0.160 g

Answers to the Practice Test

1. B	**2.** C	**3.** yes	**4.** no	**5.** yes
6. yes	**7.** no	**8.** C	**9.** E	**10.** B
11. E	**12.** D	**13.** B	**14.** C	**15.** E
16. B	**17.** C	**18.** D	**19.** D	**20.** B
21. B	**22.** E	**23.** D	**24.** A	**25.** B
26. D	**27.** A	**28.** E	**29.** A	**30.** B
31. D	**32.** B	**33.** D	**34.** C	**35.** E
36. C	**37.** A	**38.** E	**39.** D	**40.** E
41. B	**42.** A			

Study Goals

- Classify an example of matter as a pure substance or a mixture.
- Describe some physical and chemical properties of matter.
- Identify the states of matter as solids, liquids, or gases.
- Describe potential and kinetic energy.
- Identify calorie and joule as units of energy and convert between them.
- Calculate temperature values in degrees Celsius, degrees Fahrenheit, and kelvins.
- Calculate the number of joules or calories lost or gained by a specific mass of a substance for a given temperature change.
- Use the energy values for foods to calculate the kilocalories or kilojoules in a food sample.
- Describe the changes of state for solids, liquids, and gases.
- Draw a heating or cooling curve for a substance given its melting and boiling points.
- Determine the energy lost or gained during a change of state at the melting or boiling point.

Think About It

1. What kinds of activities did you do today that used *kinetic* energy?

2. Why is the energy in your breakfast cereal *potential* energy?

3. Why is the high specific heat of water important to our survival?

4. How does perspiring during a workout help to keep you cool?

5. Why is a steam burn much more damaging to skin than a hot water burn?

Key Terms

Match the following terms with the statements:

a. potential energy   **b.** joule   **c.** matter
d. chemical change   **e.** physical change   **f.** absolute zero

1. _____ the SI unit of energy

2. _____ anything that has mass and occupies space

3. _____ the lowest temperature possible, which is 0 on the Kelvin scale

4. _____ a change in shape or state in which an original substance retains its identity

5. _____ type of energy related to position or composition of a substance

6. _____ a change in which an original substance is converted into a new substance with new properties

Answers   **1.** b   **2.** c   **3.** f   **4.** e   **5.** a   **6.** d

2.1 Classification of Matter

- Matter is anything that has mass and occupies space.
- A pure substance, element or compound, has a definite composition.
- Elements are the simplest type of matter; compounds consist of a combination of two or more elements.
- Mixtures contain two or more substances that are physically, not chemically, combined.
- Mixtures are classified as homogeneous or heterogeneous.

(MC)

Tutorial: Classification of Matter
Tutorial: Classifying Matter

♦ **Learning Exercise 2.1A**

Identify each of the following as an element (E) or compound (C):

1. _____ iron   **2.** _____ carbon dioxide

3. _____ potassium iodide   **4.** _____ gold

5. _____ aluminum   **6.** _____ table salt (sodium chloride)

Answers   **1.** E   **2.** C   **3.** C   **4.** E   **5.** E   **6.** C

♦ **Learning Exercise 2.1B**

Identify each of the following as a pure substance (P) or mixture (M):

1. _____ bananas and milk   **2.** _____ sulfur

3. _____ silver   **4.** _____ a bag of raisins and nuts

5. _____ water   **6.** _____ sand and water

Answers   **1.** M   **2.** P   **3.** P   **4.** M   **5.** P   **6.** M

♦ **Learning Exercise 2.1C**

Identify each of the following mixtures as homogeneous (Ho) or heterogeneous (He):

1. _____ chocolate milk   2. _____ sand and water

3. _____ lemonade with lemon slices   4. _____ a bag of raisins and nuts

5. _____ air   6. _____ vinegar

Answers   1. Ho   2. He   3. He   4. He   5. Ho   6. Ho

2.2 States and Properties of Matter

- The three states of matter are solid, liquid, and gas.
- Physical properties are those characteristics of a substance that change without affecting the identity of a substance.
- A substance undergoes a physical change when its shape, size, or state changes, but the type of substance itself does not change.
- Chemical properties are those characteristics of a substance that change when a new substance is produced.
- In a chemical change, the original substance is converted into one or more new substances with different physical and chemical properties.

MC

Tutorial: Properties and Changes of Matter

♦ **Learning Exercise 2.2A**

State whether each of the following statements describes a gas (G), a liquid (L), or a solid (S):

1. _____ There are no attractions among the molecules.

2. _____ Particles are held close together in a definite pattern.

3. _____ The substance has a definite volume, but no definite shape.

4. _____ The particles are moving extremely fast.

5. _____ This substance has no definite shape and no definite volume.

6. _____ The particles are very far apart.

7. _____ This substance has a definite volume, but takes the shape of its container.

8. _____ The particles of this material bombard the sides of the container with great force.

9. _____ The particles in this substance are moving very, very slowly.

10. _____ This substance has a definite volume and a definite shape.

Answers   1. G   2. S   3. L   4. G   5. G
              6. G   7. L   8. G   9. S   10. S

♦ **Learning Exercise 2.2B**

Classify each of the following as a physical (P) or chemical (C) property:

1. _____ Silver is shiny.   2. _____ Water is a liquid at 25 °C.

3. _____ Wood burns.   4. _____ Mercury is a very dense liquid.

5. _____ Helium is not reactive.   6. _____ Ice cubes float in water.

Answers   **1.** P   **2.** P   **3.** C   **4.** P   **5.** C   **6.** P

♦ **Learning Exercise 2.2C**

Classify each of the following as a physical (P) or chemical change (C).

1. _____ Sodium melts at 98 °C.

2. _____ Iron forms rust in air and water.

3. _____ Water condenses on a cold window.

4. _____ Fireworks explode when ignited.

5. _____ Gasoline burns in a car engine.

6. _____ Paper is cut to make confetti.

Answers   **1.** P   **2.** C   **3.** P   **4.** C   **5.** C   **6.** P

♦ **Learning Exercise 2.2D**

Gold has been highly valued from ancient times. Gold, which has a density of 19.3 g/mL, melts at 1064 °C. It is found free in nature usually along with quartz or pyrite deposits. Gold is a shiny metal with a yellow color. It is a soft metal that is also the most malleable metal, which accounts for its use in jewelry and gold leaf. Gold is also a good conductor of heat and electricity. List the physical properties that describe gold.

Answer   density 19.3 g/mL   melting point 1064 °C
              shiny   soft
              yellow color   good conductor of heat and electricity
              malleable

2.3 Energy

- Energy is the ability to do work.
- Potential energy is energy determined by position or composition; kinetic energy is the energy of motion.
- The SI unit of energy is the joule (J), the metric unit is the calorie (cal). One calorie (cal) is equal to exactly 4.184 joules (J).

$$\frac{4.184 \text{ J}}{1 \text{ cal}} \quad \text{and} \quad \frac{1 \text{ cal}}{4.184 \text{ J}}$$

(MC)

Tutorial: Heat
Tutorial: Energy Conversions

♦ **Learning Exercise 2.3A**

Match the words in column **A** with the descriptions in column **B**.

A	B
1. ____ kinetic energy	**a.** energy due to position or composition
2. ____ potential energy	**b.** the ability to do work
3. ____ energy	**c.** the energy of motion

Answers   **1.** c   **2.** a   **3.** b

♦ **Learning Exercise 2.3B**

State whether the following statements describe potential (P) or kinetic (K) energy:

1. ____ a potted plant sitting on a ledge   2. ____ your breakfast cereal

3. ____ logs sitting in a fireplace   4. ____ a piece of candy

5. ____ an arrow shot from a bow   6. ____ a ski jumper at the top of the ski jump

7. ____ a jogger running   8. ____ a sky diver waiting to jump

9. ____ water flowing down a stream   10. ____ a bowling ball striking the pins

Answers	1. P	2. P	3. P	4. P	5. K
	6. P	7. K	8. P	9. K	10. K

♦ **Learning Exercise 2.3C**

Match the words in column **A** with the descriptions in column **B**.

A	**B**
1. ____ calorie	**a.** the SI unit of heat
2. ____ kilocalorie	**b.** the heat needed to raise 1 g of water by 1 °C
3. ____ joule	**c.** 1000 cal

Answers	1. b	2. c	3. a

♦ **Learning Exercise 2.3D**

Calculate an answer for each of the following conversions:

1. 58 000 cal to kcal   2. 3450 J to cal

3. 2.8 kJ to cal   4. 15 200 cal to kJ

Answers	1. 58 kcal	2. 825 cal	3. 670 cal	4. 63.6 kJ

2.4 Temperature

- Temperature is measured in degrees Celsius (°C), or kelvins, K. In the United States, the Fahrenheit scale (°F) is still in use.
- The equation $T_F = 1.8 T_C + 32$ is used to convert a Celsius temperature to a Fahrenheit temperature. When rearranged for T_C, the equation is used to convert a Fahrenheit temperature to a Celsius temperature.

$$T_C = \frac{(T_F - 32)}{1.8}$$

- The temperature on the Celsius scale is related to the Kelvin scale: $T_K = T_C + 273$

Tutorial: Temperature Conversions

♦ **Learning Exercise 2.4**

Calculate the temperatures in the following problems:

a. To prepare yogurt, milk is warmed to 68 °C. What Fahrenheit temperature is needed to prepare the yogurt?

b. On a chilly day in Alaska, the temperature drops to –12 °C. What is that temperature on the Kelvin scale?

c. A patient has a temperature of 39.5 °C. What is that temperature in °F?

d. On a hot summer day, the temperature is 95 °F. What is the temperature on the Celsius scale?

e. A pizza is cooked at a temperature of 425 °F. What is the °C temperature?

f. A research experiment requires the use of liquid nitrogen to cool the reaction flask to –45 °C. What temperature will this be on the Kelvin scale?

Answers	**a.** 154 °F	**b.** 261 K	**c.** 103.1 °F
	d. 35 °C	**e.** 218 °C	**f.** 228 K

2.5 Specific Heat

- Specific heat is the amount of energy required to raise the temperature of 1 g of a substance by 1 °C.
- The specific heat for liquid water is 1.00 cal/g °C or 4.184 J/g °C.
- The heat lost or gained can be calculated using the mass of the substance, temperature difference (ΔT), and its specific heat (*SH*): Heat = mass $\times$ ΔT $\times$ *SH*.

Tutorial: Heat Capacity
Tutorial: Specific Heat Calculations

Guide to Calculations Using Specific Heat	
Step 1	List given and needed data.
Step 2	Calculate the temperature change ΔT.
Step 3	Write the heat equation and rearrange for unknown.
Step 4	Substitute the given values and solve, making sure units cancel.

♦ **Learning Exercise 2.5**

Calculate the heat gained or released during the following:

1. joules and calories when 20.0 g of water is heated from 22 °C to 77 °C

2. joules and calories when 10.0 g of water is heated from 12.4 °C to 67.5 °C

3. kilojoules and kilocalories when 4.50 kg of water cools from 80.0 °C to 35.0 °C

4. kilojoules and kilocalories when 125 g of water cools from 45.0 °C to 72.0 °C

Answers   **1.** 4600 J; 1100 cal   **2.** 2310 J; 551 cal
              3. 847 kJ; 203 kcal   **4.** 14.1 kJ; 3.38 kcal

2.6 Energy and Nutrition

• A nutritional Calorie (Cal) is the same amount of energy as 1 kcal, or 1000 cal.
• When a substance is burned in a calorimeter, the water that surrounds the reaction chamber absorbs the heat given off. The heat absorbed by the water is calculated and the caloric value (energy per gram) is determined for the substance.

(MC)

Tutorial: Nutritional Energy
Case Study: Calories from Hidden Sugar

Study Note

The caloric content of a food is the sum of calories from carbohydrate, fat, and protein. It is calculated by using the number of grams of each in a food and the caloric values of 17 kJ/g (4 kcal/g) for carbohydrate and protein, and 38 kJ/g (9 kcal/g) for fat.

♦ **Learning Exercise 2.6A**

Calculate the kilojoules (kJ) and kilocalories (kcal) for the following foods using the following data: (Round off the final answer to the nearest 10 kJ or kcal)

Food	Carbohydrate	Fat	Protein	kJ	kcal
1. green peas, cooked, 1 cup	19 g	1 g	9 g		
2. potato chips, 10 chips	10 g	8 g	1 g		
3. cream cheese, 8 oz	5 g	86 g	18 g		
4. lean hamburger, 3 oz	0	10 g	23 g		
5. banana, 1	26 g	0	1 g		

Answers   **1.** 510 kJ; 120 kcal   **2.** 490 kJ; 120 kcal   **3.** 3660 kJ; 870 kcal
              4. 770 kJ; 180 kcal   **5.** 460 kJ; 110 kcal

♦ **Learning Exercise 2.6B**

Using caloric values, give answers for each of the following problems:

1. How many kcal are in a serving of pudding that contains 5 g protein, 31 g of carbohydrate, and 5 g of fat (round off kilocalories to the tens place)?

2. One serving of peanut butter (2 tbsp) has a caloric value of 190 kcal. If there are 8 g of protein and 10 g of carbohydrate, how many grams of fat are in one serving of peanut butter?

3. A serving of breakfast cereal provides 220 kcal. In this serving, there are 8 g of protein and 6 g of fat. How many grams of carbohydrate are in the cereal?

4. Complete the following table listing ingredients for a peanut butter sandwich (tbsp = tablespoon; tsp = teaspoon) (round off kilocalories to the tens place):

	Protein	Carbohydrate	Fat	kcal
2 slices of bread	5 g	30 g	0 g	
2 tbsp of peanut butter	8 g	10 g	13 g	
2 tsp of jelly	0 g	10 g	0 g	
1 tsp of margarine	0 g	0 g	7 g	
	Total kcal			

Answers

1. protein = 20 kcal; carbohydrate = 120 kcal; fat = 50 kcal Total = 190 kcal

2. protein = 30 kcal; carbohydrate = 40 kcal Then 190 kcal – 30 kcal – 40 kcal = 120 kcal of fat
   Converting 120 kcal of fat to g of fat: 120 kcal fat (1g fat/9 kcal fat) = 13 g of fat

3. protein = 30 kcal; fat = 50 kcal = 80 kcal from protein and fat

   220 kcal – 80 kcal = 140 kcal due to carbohydrate

   140 kcal (1 g carbohydrate/4 kcal) = 35 g of carbohydrate

4. bread = 140 kcal; peanut butter = 190 kcal; jelly = 40 kcal; margarine = 60 kcal
   total kcal in sandwich = 430 kcal (4.3×10^2 kcal)

2.7 Changes of State

- A substance melts and freezes at its melting (freezing) point. During the process of melting or freezing, the temperature remains constant.
- The *heat of fusion* is the energy required to change 1 g of solid to liquid. For water to freeze at 0 °C, the heat of fusion is 80. cal or 334 J/g. This is also the amount of heat lost when 1 gram of water freezes at 0 °C.

- Evaporation is a surface phenomenon while boiling occurs throughout the liquid.
- A substance boils and condenses at its boiling point. During the process of boiling or condensing, the temperature remains constant.
- When water boils at 100 °C, 540 cal or 2260 J, the *heat of vaporization* is required to change 1 g of liquid to gas (steam); it is also the amount of heat released when 1 g of water vapor condenses at 100 °C and is called the *heat of condensation*.
- Sublimation is the change of state from a solid directly to a gas. Deposition is the opposite process.
- A heating or cooling curve illustrates the changes in temperature and states as heat is added to or removed from a substance.
- The heat given off or gained when a substance undergoes temperature changes, as well as changes of state, is the total of two or more energy calculations.

(MC)

Tutorial: Heat of Vaporization and Heat of Fusion
Tutorial: Heat, Energy, and Changes of State

♦ **Learning Exercise 2.7A**

Identify each of the following as

a. melting   **b.** freezing   **c.** sublimation   **d.** condensation

1. _____ A liquid changes to a solid.

2. _____ Ice forms on the surface of a lake in winter.

3. _____ Dry ice in an ice cream cart changes to a gas.

4. _____ Butter in a hot pan turns to liquid.

5. _____ A gas changes to a liquid.

Answers   1. b   2. b   3. c   4. a   5. d

♦ **Learning Exercise 2.7B**

Calculate each of the following when a substance melts or freezes:

1. calories needed to melt 15 g ice at 0 °C

2. kilocalories released when 325 g of water freezes at 0 °C

3. grams of ice that melt when 1680 J of heat are absorbed

Answers   1. 1200 cal   2. 26 kcal   3. 5.03 g

♦ **Learning Exercise 2.7C**

Calculate each of the following when a substance boils or condenses:

1. calories needed to completely change 10. g of water to vapor at 100 °C

2. kilocalories released when 515 g of steam at 100 °C condense to form liquid water at 100 °C

3. grams of water that can be converted to steam at 100 °C when 114 kJ are absorbed

Answers   **1.** 5400 cal   **2.** 278 kcal   **3.** 50.4 g

♦ **Learning Exercise 2.7D**

On each heating or cooling curve, indicate the portion that corresponds to a solid, liquid, gas, and the changes of state.

1. Draw a heating curve for water that begins at –20 °C and ends at 120 °C. Water has a melting point of 0 °C and a boiling point of 100 °C.

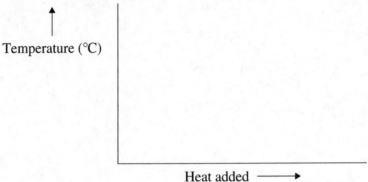

2. Draw a heating curve for bromine from –25 °C to 75 °C. Bromine has a melting point of –7 °C and a boiling point of 59 °C.

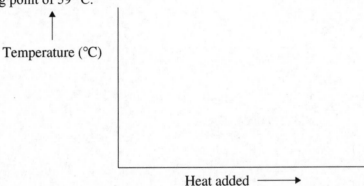

3. Draw a cooling curve for sodium from 1000 °C to 0 °C. Sodium has a freezing point of 98 °C and a boiling (condensation) point of 883 °C.

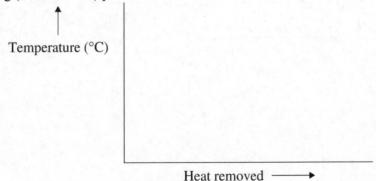

Answers:

1.

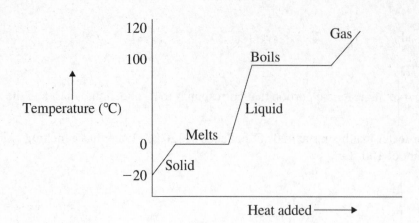

2.

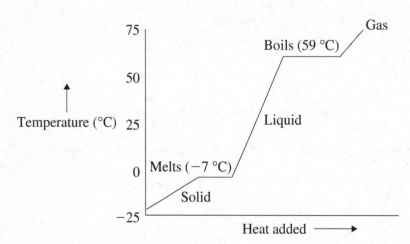

3.

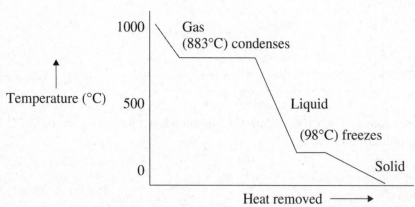

Checklist for Chapter 2

You are ready to take the Practice Test for Chapter 2. Be sure that you have accomplished the following learning goals for this chapter. If you are not sure, review the section listed at the end of the goal. Then apply your new skills and understanding to the Practice Test.

After studying Chapter 2, I can successfully:

____ Identify matter as a pure substance or mixture (2.1).

____ Identify a mixture as homogeneous or heterogeneous (2.1).

____ Identify the physical state of a substance as a solid, liquid, or gas (2.2).

____ Distinguish between a physical and chemical property (2.2).

____ Describe some forms of energy (2.3).

____ Change a quantity in one energy unit to another energy unit (2.3).

____ Given a temperature, calculate a corresponding temperature on another temperature scale (2.4).

____ Given the mass of a sample, specific heat, and the temperature change, calculate the heat lost or gained (2.5).

____ Use energy values to calculate the kilocalories or kilojoules in a food sample (2.6).

____ Calculate the heat change for the melting, freezing, boiling, and condensation of a specific quantity of a substance (2.7).

____ Draw heating and cooling curves using the melting and boiling points of a substance (2.7).

Practice Test for Chapter 2

For questions 1 through 4, classify each of the following as a pure substance (P) or a mixture (M):

1. toothpaste _____

2. platinum _____

3. chromium _____

4. mouthwash _____

For questions 5 through 8, classify each of the following mixtures as homogeneous (Ho) or heterogeneous (He):

5. noodle soup _____

6. vitamin water _____

7. chocolate chip cookie _____

8. mouthwash _____

9. Which of the following is a chemical property?
   A. dynamite explodes   **B.** a shiny metal   **C.** a melting point of 110 °C
   D. rain on a cool day   **E.** breaking up cement

For questions 10 through 12, answer with solid (S), liquid (L), or gas (G):

10. ____ has a definite volume, but takes the shape of a container

11. ____ does not have a definite shape or definite volume

12. ____ has a definite shape and a definite volume

13. Which of the following is a chemical property of silver?
    A. density of 10.5 g/mL
    B. shiny
    C. melts at 961 °C
    D. good conductor of heat
    E. reacts to form tarnish

14. Which of the following is a physical property of silicon?
    A. burns in chlorine
    B. has a black to gray color
    C. reacts with nitric acid
    D. reacts with oxygen to form sand
    E. used to form silicone

For problems 15 through 19, answer as physical change (P) or chemical change (C):

15. _____ butter melts in a hot pan

16. _____ iron forms rust with oxygen

17. _____ baking powder form bubbles (CO_2) as a cake is baking

18. _____ water boils

19. _____ propane burns in a camp stove

20. Which of the following would be described as potential energy?
    A. a car going around a racetrack
    B. a rabbit hopping
    C. oil in an oil well
    D. a moving merry-go-round
    E. a bouncing ball

21. Which of the following would be described as kinetic energy?
    A. a car battery
    B. a can of tennis balls
    C. gasoline in a car fuel tank
    D. a box of matches
    E. a tennis ball crossing over the net

22. 105 °F = _____ °C
    A. 73 °C
    B. 41 °C
    C. 58 °C
    D. 90 °C
    E. 189 °C

23. The melting point of gold is 1064 °C. The Fahrenheit temperature needed to melt gold would be
    A. 129 °C
    B. 623 °F
    C. 1031 °F
    D. 1913 °F
    E. 1947 °F

24. The average daytime temperature on the planet Mercury is 683 K. What is this temperature on the Celsius scale?
    A. 956 °C
    B. 715 °C
    C. 680 °C
    D. 410 °C
    E. 303 °C

25. Which of the following describes a liquid?
    A. A substance that has no definite shape and no definite volume.
    B. A substance with particles that are far apart.
    C. A substance with a definite shape and a definite volume.
    D. A substance containing particles that are moving very fast.
    E. A substance that has a definite volume, but takes the shape of its container.

26. The number of calories needed to raise the temperature of 5.0 g water from 25 °C to 55 °C is
    A. 5 cal
    B. 30 cal
    C. 50 cal
    D. 80 cal
    E. 150 cal

27. The number of joules needed to raise the temperature of 12.0 g gold (*SH* 0.129 J/g °C) from 40. °C to 250. °C is
   A. 1.55 J   **B.** 325 J   **C.** 480. J   **D.** 2520 J   **E.** 3000 J

28. The number of kilocalories released when 150 g of water cools from 58 °C to 22 °C is
   A. 1.1 kcal   **B.** 4.2 kcal   **C.** 5.4 kcal   **D.** 6.9 kcal   **E.** 8.7 kcal

For questions 29 through 31, consider a cup of milk with a caloric value of 165 kcal. In the cup of milk, there are 9 g of fat, 12 g of carbohydrate, and some protein.

29. The number of kcal provided by the carbohydrate is
   A. 4 kcal   **B.** 9 kcal   **C.** 36 kcal   **D.** 48 kcal   **E.** 81 kcal

30. The number of kcal provided by the fat is
   A. 4 kcal   **B.** 9 kcal   **C.** 36 kcal   **D.** 48 kcal   **E.** 81 kcal

31. The number of kcal provided by the protein is
   A. 4 kcal   **B.** 9 kcal   **C.** 36 kcal   **D.** 48 kcal   **E.** 81 kcal

For questions 32 through 35, match the term with its definition.
   A. evaporation   **B.** heat of fusion
   C. heat of vaporization   **D.** boiling

32. ____ The energy required to convert a gram of solid to liquid.

33. ____ The heat needed to boil a liquid.

34. ____ The conversion of liquid molecules to gas at the surface of a liquid.

35. ____ The formation of a gas within the liquid as well as on the surface.

36. ____ Ice cools down a drink because
   A. the ice is colder than the drink and heat flows into the ice cubes.
   B. heat is absorbed from the drink to melt the ice cubes.
   C. the heat of fusion of the ice is higher than the heat of fusion for water.
   D. Both A and B
   E. None of the above

37. The number of kilojoules released when 14.2 g of water freezes at 0 °C is
   A. 0 kJ   **B.** 1.14 kJ   **C.** 4.74 kJ   **D.** 1140 kJ   **E.** 4740 kJ

38. The number of kilocalories needed to convert 450 g of ice to liquid at 0 °C is
   A. 0 kcal   **B.** 360 kcal   **C.** 80 kcal   **D.** 240 kcal   **E.** 36 kcal

For questions 39 through 42, consider the heating curve below for p-toluidine. Answer the following questions when heat is added to p-toluidine at 20 °C where toluidine is below its melting point.

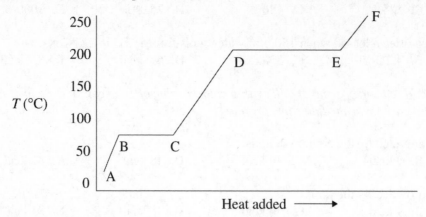

39. On the heating curve, segment BC indicates
    A. solid   **B.** melting   **C.** liquid   **D.** boiling   **E.** gas

40. On the heating curve, segment CD shows toluidine as
    A. solid   **B.** melting   **C.** liquid   **D.** boiling   **E.** gas

41. The boiling point of toluidine would be
    A. 20 °C   **B.** 45 °C   **C.** 100 °C   **D.** 200 °C   **E.** 250 °C

42. On the heating curve, segment EF shows toluidine as
    A. solid   **B.** melting   **C.** liquid   **D.** boiling   **E.** gas

Answers to the Practice Test

1. M	**2.** P	**3.** P	**4.** M	**5.** He
6. Ho	**7.** He	**8.** Ho	**9.** A	**10.** L
11. G	**12.** S	**13.** E	**14.** B	**15.** P
16. C	**17.** C	**18.** P	**19.** C	**20.** C
21. E	**22.** B	**23.** E	**24.** D	**25.** E
26. E	**27.** B	**28.** C	**29.** D	**30.** E
31. C	**32.** B	**33.** C	**34.** A	**35.** D
36. D	**37.** C	**38.** E	**39.** B	**40.** C
41. D	**42.** E			

3
Atoms and Elements

Study Goals

- Given the name of an element, write its correct symbol; from the symbol, write its name.
- Use the periodic table to identify the group and the period of an element.
- Classify an element as a metal, nonmetal, or metalloid.
- Describe each of the subatomic particles, proton, neutron, and electron, in terms of their location, charge, and relative mass.
- Describe Rutherford's gold-foil experiment and explain how it led to the current model of the atom.
- Use the atomic number and mass number of an atom to determine the number of protons, neutrons, and electrons in an atom.
- Identify an isotope from the number of protons and neutrons.
- Describe the relationship of isotopes to the atomic mass of an element on the periodic table.
- Use the percent abundance and mass of the isotopes in an element to calculate its atomic mass.
- Write the electron arrangements for elements 1–18 in the periodic table.
- Explain the relationship between electron arrangement, group number, and periodic law.
- Use the electron arrangement to draw electron dot symbols.
- Use the electron arrangements of atoms to explain trends in atomic size, ionization energy, and metallic character.

Think About It

1. Name some of elements you have seen today.

2. How are the symbols of the elements related to their names?

3. What are some elements that are part of your vitamins?

4. How much of the electromagnetic spectrum can we see?

5. What causes the different colors seen in fireworks?

Key Terms

Match each of the following key terms with the correct definition.

a. element   **b.** atom   **c.** atomic number   **d.** mass number
e. isotope   **f.** group   **g.** halogen

1. ____ the number of protons and neutrons in the nucleus of an atom

2. ____ the smallest particle of an element

3. ____ a primary substance that cannot be broken down into simpler substances

4. ____ an atom of an element that has a different number of neutrons than another atom of the same element

5. ____ the number of protons in an atom

6. ____ an element found in Group 7A (17)

7. ____ vertical column on the periodic table in which elements have similar properties

Answers   **1.** d   **2.** b   **3.** a   **4.** e   **5.** c
               6. g   **7.** f

3.1 Elements and Symbols

- Chemical symbols are one- or two-letter abbreviations for the names of the elements; only the first letter of a chemical symbol is a capital letter.
- Element names are derived from many sources, including mythology and from the names of planets, geographical features and famous people.

♦ Learning Exercise 3.1A

Write the symbols for each of the following elements:

1. carbon _____   2. iron _____   3. sodium _____

4. phosphorus _____   5. oxygen _____   6. nitrogen _____

7. iodine _____   8. sulfur _____   9. potassium _____

10. lead _____   11. calcium _____   12. gold _____

13. copper _____   14. neon _____   15. chlorine _____

Answers   **1.** C   **2.** Fe   **3.** Na   **4.** P   **5.** O
               6. N   **7.** I   **8.** S   **9.** K   **10.** Pb
               11. Ca   **12.** Au   **13.** Cu   **14.** Ne   **15.** Cl

♦ Learning Exercise 3.1B

Write the names of the elements represented by each of the following symbols:

1. Mg _____   2. H _____

3. Ag _____   4. F _____

5. Cr _____   6. Be _____

7. Si _____   8. Br _____

9. Zn _____   10. Al _____

11. Ba _____   12. Li _____

Answers
1. magnesium   2. hydrogen   3. silver
4. fluorine   5. chromium   6. beryllium
7. silicon   8. bromine   9. zinc
10. aluminum   11. barium   12. lithium

3.2 The Periodic Table

- The periodic table is an arrangement of the elements into vertical columns and horizontal rows.
- Each vertical column contains a group of elements that have similar properties.
- A horizontal row of elements is called a period.
- On the periodic table, the metals are located on the left of the heavy zigzag line, the nonmetals are to the right, and the metalloids are along the zigzag line.

Tutorial: Elements and Symbols in the Periodic Table
Tutorial: Metals, Nonmetals, and Metalloids

Study Note

1. The periodic table consists of horizontal rows called *periods* and vertical columns called *groups*, also called *families*.
2. Group 1A (1) contains the *alkali metals*, Group 2A (2) contains the *alkaline earth metals*, Group 7A (17) contains the *halogens*, and Group 8A (18) contains the *noble gases*.

♦ Learning Exercise 3.2A

Indicate whether the following elements are in a group (G), period (P), or neither (N):

1. Li, C, and O _____   2. Br, Cl, and F _____

3. Al, Si, and Cl _____   4. C, N, and O _____

5. Mg, Ca, and Ba _____   6. C, S, and Br _____

7. Li, Na, and K _____   8. K, Ca, and Br _____

Answers
1. P   2. G   3. P   4. P
5. G   6. N   7. G   8. P

♦ **Learning Exercise 3.2B**

Complete the list of elements, group numbers, and period numbers in the following table:

Element and Symbol	Group Number	Period Number
	2A (2)	3
Silicon, Si		
	5A (15)	2
Aluminum, Al		
	4A (14)	5
	1A (1)	6

Answers

Element and Symbol	Group Number	Period Number
Magnesium, Mg	2A (2)	3
Silicon, Si	4A (14)	3
Nitrogen, N	5A (15)	2
Aluminum, Al	3A (13)	3
Tin, Sn	4A (14)	5
Cesium, Cs	1A (1)	6

♦ **Learning Exercise 3.2C**

Identify each of the following elements as a metal (M), nonmetal (NM), or metalloid (ML):

1. Cl _____   2. N _____   3. Fe _____   4. Si _____   5. Al _____

6. C _____   7. Ca _____   8. Zn _____   9. Sb _____   10. Mg _____

Answers   1. NM   2. NM   3. M   4. ML   5. M
          6. NM   7. M   8. M   9. ML   10. M

♦ **Learning Exercise 3.2D**

Match the name of the group on the periodic table with one or more of the following elements: K, Cl, He, Fe, Mg, Ne, Li, Cu, and Br.

1. halogen _____

2. noble gas _____

3. alkali metal _____

4. alkaline earth metal _____

5. transition elements _____

Answers   **1.** Cl, Br   **2.** He, Ne   **3.** K, Li   **4.** Mg   **5.** Fe, Cu

3.3 The Atom

- An atom is the smallest particle that retains the characteristics of an element.
- Atoms are composed of three subatomic particles. Protons have a positive charge (+), electrons carry a negative charge (–), and neutrons are electrically neutral.
- The protons and neutrons, each with a mass of about 1 amu, are found in the tiny, dense nucleus.
- The electrons are located outside the nucleus.

Dalton's Atomic Theory

1. All matter is made up of tiny particles called atoms.

2. All atoms of a given element are identical to one another and different from atoms of other elements.

3. Atoms of two or more different elements combine to form compounds. A particular compound is always made up of the same kinds of atoms and the same number of each kind of atom.

4. A chemical reaction involves the rearrangement, separation, or combination of atoms. Atoms are never created or destroyed during a chemical reaction.

(MC)

Tutorial: The Anatomy of Atoms
Tutorial: Atomic Structure and Properties of Subatomic Particles

---

♦ **Learning Exercise 3.3A**

For each of the following statements, answer True (T) if it is consistent with Dalton's atomic theory; if not, answer False (F):

1. _____ All matter is composed of atoms.

2. _____ All atoms of an element are identical.

3. _____ Atoms combine to form compounds.

4. _____ Atoms are created when compounds are formed.

Answers   **1.** T   **2.** T   **3.** T   **4.** F

♦ **Learning Exercise 3.3B**

Match the following terms with one or more of the statements below:

   a. proton   **b.** neutron   **c.** electron   **d.** nucleus

1. _____ is found in the nucleus of an atom   2. _____ has a 1– charge

3. _____ is found outside the nucleus   4. _____ has a mass of about 1 amu

5. _____ is the small, dense center of the atom   6. _____ is neutral

Answers   **1.** a and b   **2.** c   **3.** c   **4.** a and b   **5.** d   **6.** b

3.4 Atomic Number and Mass Number

- The atomic number, which can be found on the periodic table, is the number of protons in an atom of an element.
- In a neutral atom, the number of protons is equal to the number of electrons.
- The mass number is the total number of protons and neutrons in an atom.

(MC)

Tutorial: Atomic Number and Mass Number

♦ **Learning Exercise 3.4A**

Give the number of protons in each of the following neutral atoms:

1. an atom of carbon _____

2. an atom of the element with atomic number 15 _____

3. an atom with a mass number of 40 and atomic number 19 _____

4. an atom with 9 neutrons and a mass number of 19 _____

5. a neutral atom that has 18 electrons _____

Answers   **1.** 6   **2.** 15   **3.** 19   **4.** 10   **5.** 18

Study Note

In the atomic symbol for a particular atom, the mass number appears in the upper left corner and the atomic number in the lower left corner of the atomic symbol.

$$\begin{array}{l} \text{Mass number} \quad \rightarrow \\ \text{Atomic number} \quad \rightarrow \end{array} \quad {}^{32}_{16}\text{S}$$

Study Note

1. The *atomic number* is the number of protons in every atom of an element. In neutral atoms, the number of electrons equals the number of protons.

2. The mass number is the total number of neutrons and protons in the nucleus of a particular atom.

3. The number of neutrons is mass number – atomic number.

   Example: How many protons and neutrons are in the nucleus of ${}^{109}_{47}\text{Ag}$?

   Solution: The atomic number of Ag is 47. Thus, an atom of Ag has 47 protons.
   Mass number – atomic number = number of neutrons: $109 - 47 = 62$ neutrons

---

♦ **Learning Exercise 3.4B**

Give the number of neutrons in each of the following atoms:

1. has a mass number of 42 and atomic number 20 _____

2. has a mass number of 10 and 5 protons _____

3. ${}^{30}_{14}\text{Si}$ _____

4. has a mass number of 9 and atomic number 4 _____

5. has a mass number of 22 and 10 protons _____

6. a zinc atom with a mass number of 66 _____

Answers   **1.** 22   **2.** 5   **3.** 16   **4.** 5   **5.** 12   **6.** 36

♦   **Learning Exercise 3.4C**

Complete the following table for neutral atoms.

Symbol	Atomic Number	Mass Number	Number of Protons	Number of Neutrons	Number of Electrons
	12			12	
			20	22	
		55		29	
	35			45	
		35	17		
$^{120}_{50}$Sn					

Answers

Atomic Symbol	Atomic Number	Mass Number	Number of Protons	Number of Neutrons	Number of Electrons
$^{24}_{12}$Mg	12	24	12	12	12
$^{42}_{20}$Ca	20	42	20	22	20
$^{55}_{26}$Fe	26	55	26	29	26
$^{80}_{35}$Br	35	80	35	45	35
$^{35}_{17}$Cl	17	35	17	18	17
$^{120}_{50}$Sn	50	120	50	70	50

3.5 Isotopes and Atomic Mass

• Isotopes are atoms that have the same number of protons but different numbers of neutrons.
• The atomic mass of an element is the weighted average mass of all the isotopes in a naturally occurring sample of that element. The atomic mass is **not** the same as the mass number.

Self-Study Activity: Atoms and Isotopes
Tutorial: Atomic Mass Calculations

♦ **Learning Exercise 3.5A**

Identify the sets of atoms that are isotopes.

    A. $^{20}_{10}X$ **B.** $^{20}_{11}X$ **C.** $^{21}_{11}X$ **D.** $^{19}_{10}X$ **E.** $^{19}_{9}X$

Answer   Atoms A and D are isotopes (at. no. 10); atoms B and C are isotopes (at. no. 11).

♦ **Learning Exercise 3.5B**

ESSAY: Copper has two naturally occurring isotopes, $^{63}_{29}Cu$ and $^{65}_{29}Cu$. If that is the case, why is the atomic mass of copper listed as 63.55 on the periodic table?

Answer   Copper in nature consists of two isotopes with different atomic masses. The atomic mass is the average of the individual masses of the two isotopes and their percent abundance in the sample. The atomic mass does not represent the mass of any individual atom.

♦ **Learning Exercise 3.5C**

Rubidium is the element with atomic number 37.

  1. What is the symbol for rubidium?_____

  2. What is the group number for rubidium?_____

  3. What is the name of the family that includes rubidium? _____

  4. Rubidium has two naturally occurring isotopes, $^{85}_{37}Rb$ and $^{87}_{37}Rb$. How many protons, neutrons, and electrons are in each isotope?

  5. In naturally occurring rubidium, 72.17% is $^{85}_{37}Rb$ with a mass of 84.91 amu; 27.83% is $^{87}_{37}Rb$ with a mass of 86.91 amu. What is the atomic mass of rubidium?

Answers

    1. Rb
    2. Rubidium is in Group 1A (1).
    3. Rubidium is in the family called the alkali metals.
    4. The isotope $^{85}_{37}Rb$ has 37 protons, 48 neutrons, and 37 electrons. The isotope $^{87}_{37}Rb$ has 37 protons, 50 neutrons, and 37 electrons.
    5. 84.91 amu (72.17/100) + 86.91 amu (27.83/100) = 61.28 amu + 24.19 amu = 85.47 amu for the atomic mass of Rb.

3.6 Electron Energy Levels

• Electromagnetic radiation is energy that travels as waves at the speed of light.
• The electromagnetic spectrum is all forms of electromagnetic radiation, arranged in order of decreasing wavelength.

- An atomic spectrum is a series of colored lines that correspond to the specific energies emitted by a heated element.
- In an atom, the electrons of similar energy are grouped in specific energy levels. The first level can hold two electrons, the second level can hold eight electrons, and the third level will take up to 18 electrons.
- The electron arrangement specifies the number of electrons in each energy level beginning with the lowest energy level.

(MC)

Tutorial: Electromagnetic Radiation
Tutorial: Bohr's Shell Model of the Atom
Tutorial: Energy Levels

◆ **Learning Exercise 3.6**

Write the electron arrangements for the following elements:

	Electron Level		**Electron Level**
Element	1 2 3 4	**Element**	1 2 3 4
1. beryllium	_____	2. phosphorus	_____
3. carbon	_____	4. nitrogen	_____
5. potassium	_____	6. chlorine	_____
7. sodium	_____	8. silicon	_____

Answers	**1.** 2,2	**2.** 2,8,5	**3.** 2,4	**4.** 2,5	**5.** 2,8,8,1
	6. 2,8,7	**7.** 2,8,1	**8.** 2,8,4		

3.7 Trends in Periodic Properties

- The physical and chemical properties of the elements change in a periodic manner going across each period and are repeated in each successive period.
- Representative elements in a group have similar behavior.
- The group number of a representative element gives the number of valence electrons in the atoms of that group.
- An electron-dot symbol is a convenient way to represent the valence electrons, which are shown as dots on the sides, top, or bottom of the symbol for an element. For beryllium, an electron-dot symbol is:

$$\overset{\cdot}{Be} \cdot$$

- The atomic radius of representative elements generally increases going down a group and decreases going across a period.
- The ionization energy of representative elements generally decreases going down a group and increases going across a period.
- The metallic character of representative elements generally increases going down a group and decreases going across a period.

MC

Tutorial: Ionization Energy
Tutorial: Periodic Trends
Tutorial: Patterns in the Periodic Table
Tutorial: Electron-Dot Symbols for Elements

♦ **Learning Exercise 3.7A**

State the number of valence electrons, the group number of each element, and write the electron-dot symbol for each of the following:

Element	Valence Electrons	Group Number	Electron-Dot Symbol
1. sulfur			
2. oxygen			
3. magnesium			
4. hydrogen			
5. fluorine			
6. aluminum			

Answers

1. sulfur $6\,e^-$ Group 6A (16) $\cdot \ddot{\underset{\cdot\cdot}{S}} :$

2. oxygen $6\,e^-$ Group 6A (16) $\cdot \ddot{\underset{\cdot}{O}} :$

3. magnesium $2\,e^-$ Group 2A (2) $\overset{\cdot}{Mg} \cdot$

4. hydrogen $1\,e^-$ Group 1A (1) $H \cdot$

5. fluorine $7\,e^-$ Group 7A (17) $\cdot \ddot{\underset{\cdot\cdot}{F}} :$

6. aluminum $3\,e^-$ Group 3A (13) $\cdot \overset{\cdot}{Al} \cdot$

♦ **Learning Exercise 3.7B**

Indicate the element that has the larger atomic radius.

1.____ Mg or Ca   2. ____ Si or Cl

3.____ Sr or Rb   4. ____ Br or Cl

5.____ Li or Cs   6. ____ Li or N

7.____ N or P   8. ____ As or Ca

Answers   1. Ca   2. Si   3. Rb   4. Br
              5. Cs   6. Li   7. P   8. Ca

♦ **Learning Exercise 3.7C**

Indicate the element in each of the following that has the lower ionization energy:

1. ____ Mg or Na   2. ____ P or Cl

3. ____ K or Rb   4. ____ Br or F

5. ____ Li or O   6. ____ Sb or N

7. ____ K or Br   8. ____ S or Na

Answers   1. Na   2. P   3. Rb   4. Br
               5. Li   6. Sb   7. K   8. Na

♦ **Learning Exercise 3.7D**

Indicate the element that has more metallic character.

1. ____ K or Na   2. ____ O or Se

3. ____ Ca or Br   4. ____ I or F

5. ____ Li or N   6. ____ Pb or Ba

Answers   1. K   2. Se   3. Ca   4. I
               5. Li   6. Ba

Checklist for Chapter 3

You are ready to take the Practice Test for Chapter 3. Be sure that you have accomplished the following learning goals for this chapter. If you are not sure, review the section listed at the end of the goal. Then apply your new skills and understanding to the Practice Test.

After studying Chapter 3, I can successfully:

____ Write the correct symbol or name for an element (3.1).

____ Use the periodic table to identify the group and period of an element, and describe it as a metal,

nonmetal, or metalloid (3.2).

____ State the electrical charge, mass, and location of the protons, neutrons, and electrons in an atom (3.3).

____ Given the atomic number and mass number of an atom, state the number of protons, neutrons, and

electrons (3.4).

____ Identify an isotope and describe the atomic mass of an element (3.5).

____ Write the electron arrangement for the elements with atomic number 1–18 (3.6).

____ Write the electron-dot symbol for a representative element (3.7).

____ Predict the periodic trends in valence electrons, group numbers, atomic size, ionization energy, and

metallic character going down a group or across a period (3.7).

Practice Test for Chapter 3

For questions 1 through 5, write the correct symbol for each of the elements listed:

1. potassium ____   2. phosphorus ____

3. calcium ____   4. carbon ____

5. sodium ____

For questions 6 through 10, write the correct name for each of the symbols:

6. Fe _____   7. Cu _____

8. Cl _____   9. Pb _____

10. Ag _____

11. The elements C, N, and O are part of a
    A. period   B. group   C. neither

12. The elements Li, Na, and K are part of a
    A. period   B. group   C. neither

13. What is the classification of an atom with 15 protons and 17 neutrons?
    A. metal   B. nonmetal   C. transition element
    D. noble gas   E. halogen

14. What is the group number of the element with atomic number 3?
    A. 1   B. 2   C. 3   D. 7   E. 8

For questions 15 through 18, consider an atom with 12 protons and 13 neutrons.

15. This atom has an atomic number of
    A. 12   B. 13   C. 23   D. 24   E. 25

16. This atom has a mass number of
    A. 12   B. 13   C. 23   D. 24   E. 25

17. This is an atom of
    A. carbon   B. sodium   C. magnesium   D. aluminum   E. manganese

18. The number of electrons in this atom is
    A. 12   B. 13   C. 23   D. 24   E. 25

For questions 19 through 22, consider an atom of calcium with a mass number of 42.

19. This atom of calcium has an atomic number of
    A. 20   B. 22   C. 40   D. 41   E. 42

20. The number of protons in this atom of calcium is
    A. 20   B. 22   C. 40   D. 41   E. 42

21. The number of neutrons in this atom of calcium is
    A. 20   B. 22   C. 40   D. 41   E. 42

22. The number of electrons in this atom of calcium is
    A. 20   B. 22   C. 40   D. 41   E. 42

23. Platinum, $^{195}_{78}Pt$, has
    A. $78\,p^+, 78\,e^-, 78\,n$ B. $195\,p^+, 195\,e^-, 195\,n$ C. $78\,p^+, 78\,e^-, 195\,n$ D. $78\,p^+, 78\,e^-, 117\,n$
    E. $78\,p^+, 117\,e^-, 117\,n$

For questions 24 and 25, use the following list of atoms.

$$^{14}_{7}V \qquad ^{16}_{8}W \qquad ^{19}_{9}X \qquad ^{16}_{7}Y \qquad ^{18}_{8}Z$$

24. Which atoms(s) is (are) isotopes of an atom with 8 protons and 9 neutrons?
    A. W   B. W, Z   C. X, Y   D. X   E. Y

25. Which atom(s) is (are) isotopes of an atom with 7 protons and 8 neutrons?
    A. V   B. W   C. V, Y   D. W, Z   E. none

26. Which element would you expect to have properties most like oxygen?
    A. nitrogen   B. carbon   C. chlorine   D. argon   E. sulfur

27. Which of the following is an isotope of nitrogen?
    A. $^{14}_{8}N$ B. $^{7}_{3}N$ C. $^{10}_{5}N$ D. $^{4}_{2}N$ E. $^{15}_{7}N$

28. Except for helium, the number of valence electrons of the noble gases is
    A. 3   B. 5   C. 7   D. 8   E. 12

29. The electron arrangement for an oxygen atom is
    A. 2,4   B. 2,8   C. 2,6   D. 2,4,2   E. 2,6,2

30. The electron arrangement for aluminum is
    A. 2,11   B. 2,8,5   C. 2,8,3   D. 2,10,1   E. 2,2,6,3

31. The number of valence electrons in gallium is
    A. 1   B. 2   C. 3
    D. 13   E. 31

32. The atomic radius of oxygen is larger than that of
    A. lithium   B. sulfur   C. fluorine
    D. boron   E. argon

33. Of Li, Na, K, Rb, and Cs, the element with the most metallic character is
    A. Li   B. Na   C. K
    D. Rb   E. Cs

34. Of C, Si, Ge, Sn, and Pb, the element with the greatest ionization energy is
    A. C   B. Si   C. Ge
    D. Sn   E. Pb

35. The electron-dot symbol ·X· would be correct for an element in
   A. Group 1A (1)   **B.** Group 2A (2)   **C.** Group 4A (14)
   D. Group 6A (16)   **E.** Group 7A (17)

36. The electron-dot symbol X· would be correct for
   A. magnesium   **B.** chlorine   **C.** sulfur
   D. cesium   **E.** nitrogen

Answers for the Practice Test

1. K	**2.** P	**3.** Ca	**4.** C	**5.** Na
6. iron	**7.** copper	**8.** chlorine	**9.** lead	**10.** silver
11. A	**12.** B	**13.** B	**14.** A	**15.** A
16. E	**17.** C	**18.** A	**19.** A	**20.** A
21. B	**22.** A	**23.** D	**24.** B	**25.** C
26. E	**27.** E	**28.** D	**29.** C	**30.** C
31. C	**32.** C	**33.** E	**34.** A	**35.** B
36. D				

4
Compounds and Their Bonds

Study Goals

- Use the octet rule to determine the ionic charge of an ion of a representative element.
- Use charge balance to write an ionic formula.
- Write the formula and name for an ionic compound including one containing a polyatomic ion.
- Draw the electron-dot formula for a covalent compound.
- Give the name of a covalent compound from the formula.
- Use electronegativity to determine the polarity of a bond.
- Use VSEPR theory to determine the shape and bond angles of a molecule.
- Identify a covalent compound as polar or nonpolar.
- Describe the types of attractive forces that hold particles together.

Think About It

1. How does a compound differ from an element?

2. How is a covalent bond different from an ionic bond?

3. What are some compounds listed on the labels of your vitamins, toothpaste, and foods?

4. What makes salt an ionic compound?

5. Why are some covalent compounds polar and others nonpolar?

Key Terms

Match each the following key terms with the correct definition:

    a. molecule   **b.** ion   **c.** ionic bond

    d. covalent bond   **e.** octet

1. ____ a sharing of valence electrons by two atoms

2. ____ an arrangement of eight electrons in the outer energy level

3. ____ the attraction between positively and negatively charged particles

4. ____ the smallest unit of two or more atoms held together by covalent bonds

5. ____ an atom or group of atoms with a positive or negative charge

Answers   **1.** d   **2.** e   **3.** c   **4.** a   **5.** b

4.1 Octet Rule and Ions

- The stability of the noble gases is associated with eight electrons, an octet, in their outermost energy levels. Helium is stable with two electrons in its outermost energy level.

  He   2   Ne   2,8   Ar   2,8,8

  Atoms of elements other than the noble gases achieve stability by losing, gaining, or sharing their valence electrons with other atoms in the formation of compounds.

- In ionic compounds, atoms of metals in Groups 1A (1), 2A (2), and 3A (13) lose valence electrons to achieve a noble gas electron arrangement, forming positively charged cations with a charge of 1+, 2+, or 3+.

- In ionic compounds, nonmetals in Groups 5A (15), 6A (16) and 7A (17) gain valence electrons to achieve a noble gas electron arrangement, forming negatively charged anions with a charge of 3–, 2–, or 1–.

(MC)

Tutorial: Octet Rule and Ions
Tutorial: Writing Electron-Dot Formulas

Study Note

When an atom loses or gains electrons, it acquires the electron configuration of the nearest noble gas. For example, sodium loses one electron, which gives the Na^+ ion the electron configuration of neon. Oxygen gains two electrons to give an oxide ion, O^{2-}, the electron configuration of neon.

♦ **Learning Exercise 4.1A**

The following elements lose electrons when they form ions. Indicate the group number, the number of electrons lost, and the ion (symbol and charge) for each of the following:

Element	Group Number	Electrons Lost	Ion Formed
Magnesium			
Sodium			
Calcium			
Potassium			
Aluminum			

Answers

Element	Group Number	Electrons Lost	Ion Formed
Magnesium	2A (2)	2	Mg^{2+}
Sodium	1A (1)	1	Na^+
Calcium	2A (2)	2	Ca^{2+}
Potassium	1A (1)	1	K^+
Aluminum	3A (13)	3	Al^{3+}

Study Note

The valence electrons are the electrons in the outermost energy level of an atom. For representative elements, the number of valence electrons is related to the group number.

♦ **Learning Exercise 4.1B**

The following elements gain electrons when they form ions. Indicate the group number, the number of electrons gained, and the ion (symbol and charge) for each of the following:

Element	Group Number	Electrons Gained	Ion Formed
Chlorine			
Oxygen			
Nitrogen			
Fluorine			
Sulfur			

Answers

Element	Group Number	Electrons Gained	Ion Formed
Chlorine	7A (17)	1	Cl^-
Oxygen	6A (16)	2	O^{2-}
Nitrogen	5A (15)	3	N^{3-}
Fluorine	7A (17)	1	F^-
Sulfur	6A (16)	2	S^{2-}

4.2 Ionic Compounds

- In the formulas of ionic compounds, the total positive charge is equal to the total negative charge. For example, the compound magnesium chloride, $MgCl_2$, contains Mg^{2+} and two Cl^-. The sum of the charges is zero: $(2+) + 2(1-) = 0$.
- When two or more ions are needed for charge balance, that number is indicated by subscripts in the formula.

Tutorial: Ionic Compounds
Tutorial: Writing Ionic Formulas

♦ **Learning Exercise 4.2A**

For this exercise, you may want to cut pieces of paper that represent typical positive and negative ions as shown below. To determine an ionic formula, place the pieces together with the positive ion(s) on the left. Add more positive or negative ions (squares or rectangles) to complete a geometric shape. Write the number of positive and negative ions as the subscripts for the formula.

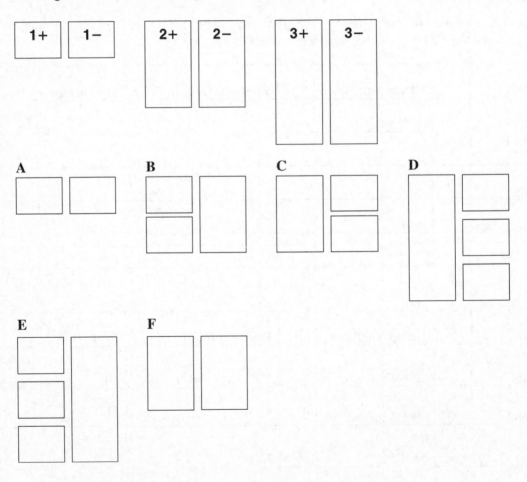

Give the letter (**A**, **B**, **C**, etc.) that matches the arrangement of ions in the following compounds:

Compound	Combination	Compound	Combination
1. $MgCl_2$	_____	2. Na_2S	_____
3. $LiCl$	_____	4. CaO	_____
5. K_3N	_____	6. $AlBr_3$	_____
7. MgS	_____	8. $BaCl_2$	_____

Answers	1. **C**	2. **B**	3. **A**	4. **F**
	5. **E**	6. **D**	7. **F**	8. **C**

Study Note

You can check that the ionic formula you write is electrically neutral by multiplying each of the ionic charges by their subscripts. When added together, their sum should equal zero. For example, the formula Na_2O gives $2(1+) + 1(2-) = (2+) + (2-) = 0$.

♦ **Learning Exercise 4.2B**

Write the correct ionic formula for the compound formed from the following pairs of ions:

1. Na^+ and Br^- _____
2. K^+ and S^{2-} _____
3. Al^{3+} and O^{2-} _____
4. Mg^{2+} and I^- _____
5. Ca^{2+} and S^{2-} _____
6. Al^{3+} and Cl^- _____
7. Li^+ and N^{3-} _____
8. Ba^{2+} and P^{3-} _____

Answers	1. $NaBr$	2. K_2S	3. Al_2O_3	4. MgI_2
	5. CaS	6. $AlCl_3$	7. Li_3N	8. Ba_3P_2

4.3 Naming and Writing Ionic Formulas

- In naming ionic compounds, the cation from a metal in Group 1A (1), 2A (2), or 3A (13) is named first followed by the name of the anion from a nonmetal obtained by using the first syllable of its elemental name followed by *ide*.
- Most transition elements form cations with two or more ionic charges. Then the ionic charge must be written as a Roman numeral after the name of the metal. For example, the cations of iron, Fe^{2+} and Fe^{3+}, are named iron(II) and iron(III). The ions of copper are Cu^+, copper(I), and Cu^{2+}, copper(II).
- The only transition elements with fixed charges are zinc, Zn^{2+}, silver, Ag^+, and cadmium, Cd^{2+}.

(MC)

Tutorial: Writing Ionic Formulas

	Guide to Naming Ionic Compounds with Metals That Form a Single Ion
Step 1	Identify the cation and anion.
Step 2	Name the cation by its element name.
Step 3	Name the anion by using the first syllable of its element name followed by *ide*.
Step 4	Write the name of the cation first and the name of the anion second.

♦ **Learning Exercise 4.3A**

Write the ions and the correct name for each of the following ionic compounds:

Formula	Ions	Name
1. Cs_2O		
2. $BaBr_2$		
3. Mg_3P_2		
4. Na_2S		

Answers
1. Cs^+, O^{2-}, cesium oxide
2. Ba^{2+}, Br^-, barium bromide
3. Mg^{2+}, P^{3-}, magnesium phosphide
4. Na^+, S^{2-}, sodium sulfide

♦ **Learning Exercise 4.3B**

Many of the transition elements form two or more ions with positive charge. Complete the following table:

Name of ion	Symbol of ion	Name of ion	Symbol of ion
1. nickel(III)	_____	2. _____	Co^{2+}
3. zinc	_____	4. _____	Fe^{2+}
5. copper(I)	_____	6. _____	Ag^+
7. gold(III)	_____	8. _____	Pb^{4+}
9. cadmium	_____	10. _____	Mn^{2+}

Answers
1. Ni^{3+} 2. cobalt(II) 3. Zn^{2+} 4. iron(II) 5. Cu^+
6. silver 7. Au^{3+} 8. lead(IV) 9. Cd^{2+} 10. manganese(II)

♦ **Learning Exercise 4.3C**

Write the name of each of the following ions:

1. Cl^- _____

2. Cr^{3+} _____

3. Au^+ _____

4. Li^+ _____

5. O^{2-} _____

6. Ca^{2+} _____

7. S⁻ _____   **8.** Al^{3+} _____

9. Co^{3+} _____   **10.** Ba^{2+} _____

11. Sn^{2+} _____   **12.** N^{3-} _____

Answers   **1.** chloride   **2.** chromium(III)   **3.** gold(I)   **4.** lithium   **5.** oxide
            6. calcium   **7.** sulfide   **8.** aluminum   **9.** cobalt(III)   **10.** barium
            11. tin(II)   **12.** nitride

	Guide to Naming Ionic Compounds with Variable Charge Metals
Step 1	Determine the charge of the cation from the anion.
Step 2	Name the cation by its element name and use a Roman numeral in parentheses for the charge.
Step 3	Name the anion by using the first syllable of its element name followed by *ide*.
Step 4	Write the name of the cation first and the name of the anion second.

♦ **Learning Exercise 4.3D**

Write the ions and a correct name for each of the following ionic compounds:

Formula	Ions		Name
1. $CrCl_2$	_____	_____	_____
2. $SnBr_4$	_____	_____	_____
3. Na_3P	_____	_____	_____
4. Ni_2O_3	_____	_____	_____
5. MnS	_____	_____	_____
6. Mg_3N_2	_____	_____	_____

Answers   **1.** Cr^{2+}, Cl^-, chromium(II) chloride   **2.** Sn^{4+}, Br^-, tin(IV) bromide
            3. Na^+, P^{3-}, sodium phosphide   **4.** Ni^{3+}, O^{2-}, nickel(III) oxide
            5. Mn^{2+}, S^{2-}, manganese(II) sulfide   **6.** Mg^{2+}, N^{3-}, magnesium nitride

	Guide to Writing Formulas from the Name of an Ionic Compound
Step 1	Identify the cation and anion.
Step 2	Balance the charges.
Step 3	Write the formula, cation first, using subscripts from the charge balance.

♦ **Learning Exercise 4.3E**

Write the ions and the correct ionic formula for the following ionic compounds:

Compound	Positive Ion	Negative Ion	Formula of Compound
Aluminum sulfide			
Tin(IV) chloride			
Magnesium oxide			
Gold(III) bromide			
Silver nitride			

Answers

Compound	Positive Ion	Negative Ion	Formula of Compound
Aluminum sulfide	Al^{3+}	S^{2-}	Al_2S_3
Tin(IV) chloride	Sn^{4+}	Cl^-	$SnCl_4$
Magnesium oxide	Mg^{2+}	O^{2-}	MgO
Gold(III) bromide	Au^{3+}	Br^-	$AuBr_3$
Silver nitride	Ag^+	N^{3-}	Ag_3N

4.4 Polyatomic Ions

- A polyatomic ion is a group of nonmetal atoms with an overall electrical charge, usually negative, 1–, 2–, or 3–. The ammonium ion, NH_4^+, has a positive charge.
- Polyatomic ions cannot exist alone but are combined with an ion of the opposite charge.
- Most ionic compounds containing three elements (polyatomic ions) have names that end with *ate* or *ite*.
- When more than one polyatomic ion is needed in the formula for a compound, the entire polyatomic formula is enclosed in parentheses and the subscript for charge balance is written outside the parentheses.

(MC)

Tutorial: Polyatomic Ions

Study Note

By learning the most common polyatomic ions such as nitrate NO_3^-, carbonate CO_3^{2-}, sulfate SO_4^{2-}, and phosphate PO_4^{3-}, you can derive their related polyatomic ions. For example, the nitrite ion, NO_2^-, has one oxygen atom fewer than the nitrate ion, NO_3^-.

♦ **Learning Exercise 4.4A**

Write the polyatomic ion (symbol and charge) for each of the following:

1. Sulfate ion _____   2. Hydroxide ion _____

3. Carbonate ion _____   4. Sulfite ion _____

5. Ammonium ion _____   6. Phosphate ion _____

7. Nitrate ion _____   8. Nitrite ion _____

Answers: 1. SO_4^{2-} 2. OH^- 3. CO_3^{2-} 4. SO_3^{2-}
             5. NH_4^+ 6. PO_4^{3-} 7. NO_3^- 8. NO_2^-

Guide to Naming Ionic Compounds with Polyatomic Ions	
Step 1	Identify the cation and polyatomic ion (anion).
Step 2	Name the cation, using a Roman numeral in parentheses, if needed.
Step 3	Name the polyatomic ion usually ending in *ite* or *ate*.
Step 4	Write the name of the compound, cation first and the polyatomic ion second.

♦ **Learning Exercise 4.4B**

Write the formula of the ions, and the correct formula for each of the following compounds:

Compound	Positive ion	Negative ion	Formula
Sodium phosphate			
Iron(II) hydroxide			
Ammonium carbonate			
Silver bicarbonate			
Chromium(III) sulfate			
Lead(II) nitrate			
Potassium sulfite			
Barium phosphate			

Compound	Positive ion	Negative ion	Formula
Sodium phosphate	Na^+	PO_4^{3-}	Na_3PO_4
Iron(II) hydroxide	Fe^{2+}	OH^-	$Fe(OH)_2$
Ammonium carbonate	NH_4^+	CO_3^{2-}	$(NH_4)_2CO_3$
Silver bicarbonate	Ag^+	HCO_3^-	$AgHCO_3$
Chromium(III) sulfate	Cr^{3+}	SO_4^{2-}	$Cr_2(SO_4)_3$
Lead(II) nitrate	Pb^{2+}	NO_3^-	$Pb(NO_3)_2$
Potassium sulfite	K^+	SO_4^{2-}	K_2SO_3
Barium phosphate	Ba^{2+}	PO_4^{3-}	$Ba_3(PO_4)_2$

4.5 Covalent Compounds and Their Names

- In a covalent bond, atoms of nonmetals share electrons to achieve an octet. For example, oxygen with six valence electrons shares electrons with two hydrogen atoms to form the covalent compound water (H_2O).

$$H \!:\! \overset{\displaystyle ..}{\underset{\displaystyle H}{O}} \!:\!$$

- In a double bond, two pairs of electrons are shared between the same two atoms. In a triple bond, three pairs of electrons are shared.
- Covalent compounds are composed of nonmetals bonded together to give discrete units called molecules.
- The formula of a covalent compound is written using the order of the symbols of the elements in the name followed by subscripts given by the prefixes.

(MC)

Self Study Activity: Covalent Bonds
Tutorial: Writing Electron-Dot Formulas
Tutorial: Covalent Lewis-Dot Structures
Tutorial: Naming Covalent Compounds

Guide to Drawing Electron-Dot Formulas	
Step 1	Determine the arrangement of atoms.
Step 2	Determine the total number of valence electrons.
Step 3	Attach each bonded atom to the central atom with a pair of electrons.
Step 4	Place the remaining electrons using single or multiple bonds to complete octets (two for H, six for B).

♦ **Learning Exercise 4.5A**

Write the electron-dot formulas for the following covalent compounds (the central atom is underlined):

1. H_2 2. $\underline{N}Cl_3$ 3. HCl

4. Cl_2 5. $H_2\underline{S}$ 6. $\underline{C}Cl_4$

Answers   1. H:H   2. :Cl:N:Cl:
            :Cl: 3. H:Cl:

4. :Cl:Cl: 5. H:S:
              H   6. :Cl:C:Cl:
                                       :Cl:

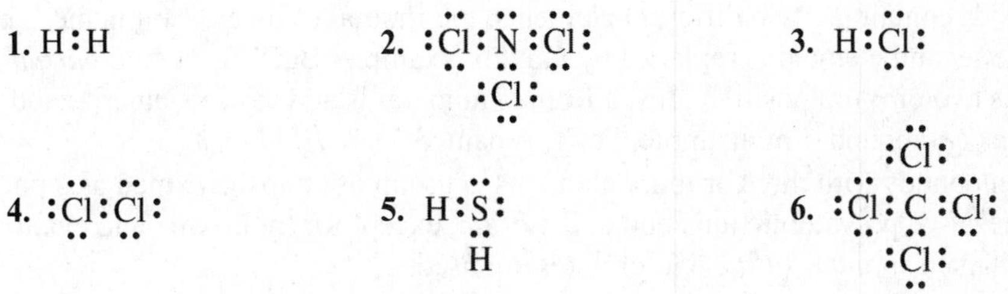

Study Note

Two nonmetals can form two or more different covalent compounds. In their names, prefixes are used to indicate the subscript in the formula. Some typical prefixes are mono (1), di (2), tri (3), tetra (4), and penta (5). The ending of the second nonmetal is changed to *-ide*.

♦ **Learning Exercise 4.5B**

Use the appropriate prefixes in naming the following covalent compounds:

1. CS_2 _____   2. SO_2 _____

3. CCl_4 _____   4. N_2O_4 _____

5. CO _____   6. PCl_3 _____

Answers   1. carbon disulfide   2. sulfur dioxide   3. carbon tetrachloride
            4. dinitrogen tetroxide   5. carbon monoxide   6. phosphorus trichloride

---

Guide to Writing Formulas for Covalent Compounds	
Step 1	Write the symbols in the order of the elements in the name.
Step 2	Write any prefixes as subscripts.

♦ **Learning Exercise 4.5C**

Write the formula of each of the following covalent compounds:

1. dinitrogen oxide _____ 2. carbon dioxide _____

3. silicon tetrabromide _____ 4. sulfur hexafluoride _____

5. nitrogen trichloride _____ 6. oxygen difluoride _____

Answers 1. N_2O 2. CO_2 3. $SiBr_4$ 4. SF_6 5. NCl_3 6. OF_2

Summary of Writing Formulas and Names

- In compounds containing two different elements, the first takes its element name. The ending of the name of the second element is replaced by *ide*. For example, $BaCl_2$ is named *barium chloride.* If the metal forms two or more positive ions, a Roman numeral is added to its name to indicate the ionic charge in the compound. For example, $FeCl_3$ is named *iron(III) chloride.*
- In ionic compounds with three or more elements, a group of atoms is named as a polyatomic ion. The names of negative polyatomic ions end in *ate* or *ite*, except for hydroxide and cyanide. For example, Na_2SO_4 is named *sodium sulfate.* No prefixes are used.
- When a polyatomic ion occurs two or more times in a formula, its formula is placed inside parentheses, and the number of ions are shown as a subscript after the parentheses $Ca(NO_3)_2$.
- In naming covalent compounds, a prefix before the name of an element indicates the numerical value of a subscript. For example, N_2O_3 is named *dinitrogen trioxide.*

♦ **Learning Exercise 4.5D**

Indicate the type of compound (ionic or covalent) formed from each pair of elements. If it is ionic, write the ions; if covalent, write the electron-dot formula of the molecule. Then give a formula and name for each.

Components	Ionic or Covalent?	Ions or Electron-Dot Formula	Formula	Name
Mg and Cl				
N and Cl				
K and SO_4				
Li and O				
C and Cl				
Na and PO_4				
H and S				
Ca and HCO_3				

Answers

Mg and Cl	Ionic	Mg^{2+}, Cl^-	$MgCl_2$	Magnesium chloride
N and Cl	Covalent	:Cl:N:Cl: :Cl:	NCl_3	Nitrogen trichloride
K and SO_4	Ionic	K^+, SO_4^{2-}	K_2SO_4	Potassium sulfate
Li and O	Ionic	Li^+, O^{2-}	Li_2O	Lithium oxide
C and Cl	Covalent	:Cl: :Cl:C:Cl: :Cl:	CCl_4	Carbon tetrachloride
Na and PO_4	Ionic	Na^+, PO_4^{3-}	Na_3PO_4	Sodium phosphate
H and S	Covalent	H:S: H	H_2S	Dihydrogen sulfide
Ca and HCO_3	Ionic	Ca^{2+}, HCO_3^-	$Ca(HCO_3)_2$	Calcium hydrogen carbonate (bicarbonate)

4.6 Electronegativity and Bond Polarity

- Electronegativity values indicate the ability of an atom to attract electrons in a chemical bond. In general, metals have low electronegativity values and nonmetals have high values.
- When atoms sharing electrons have the same or similar electronegativity values, electrons are shared equally and the bond is *nonpolar covalent.*
- Electrons are shared unequally in *polar covalent* bonds because they are attracted to the more electronegative atom.
- An electronegativity difference of 0 to 0.4 indicates a nonpolar covalent bond, while a difference of 0.5 to 1.8 indicates a polar covalent bond.
- An electronegativity difference greater than 1.8 indicates a bond that is *ionic.*
- A polar bond with its charge separation is called a dipole; the positive end is marked as δ^+ and the negative end as δ^- .

(MC)

Tutorial: Electronegativity
Self Study Activity: Bonds and Bond Polarities

♦ **Learning Exercise 4.6A**

Using the electronegativity values, determine the electronegativity difference for each bond and the type of bonding: (I) ionic, (PC) polar covalent, or (NC) nonpolar covalent.

Elements	Electronegativity difference	Bonding	Elements	Electronegativity difference	Bonding
a. H—O	_____	_____	b. H—Cl	_____	_____
c. N—S	_____	_____	d. Cl—Cl	_____	_____
e. Al—O	_____	_____	f. S—F	_____	_____
g. Li—F	_____	_____	h. H—C	_____	_____

Answers   a. 1.4, PC   b. 0.9, PC   c. 0.5, PC   d. 0.0, NP
             e. 2.0, I   f. 1.5, PC   g. 3.0, I   h. 0.4, NP

♦ **Learning Exercise 4.6B**

Write the symbols δ^+ and δ^- over the atoms in the polar covalent bonds or label nonpolar covalent.

1. H—O   2. N—N   3. C—Cl

4. O—F   5. N—F   6. Cl—Cl

Answers   1. δ^+ δ^- 2. nonpolar covalent   3. δ^+ δ^-
                H — O   C — Cl

             4. δ^+ δ^- 5. δ^+ δ^- 6. nonpolar covalent
                O — F   N — F

4.7 Shapes and Polarity of Molecules

- Valence shell electron-pair repulsion (VSEPR) theory predicts the geometry of a molecule by placing the electron groups around a central atom as far apart as possible.
- A central atom with two electron groups has a linear geometry (180°); three electron groups give a trigonal planar geometry (120°); and four electron groups give a tetrahedral geometry (109°).
- A linear molecule has a central atom bonded to two atoms and no lone pairs.
- A trigonal planar molecule has a central atom bonded to three atoms and no lone pairs, with bond angles of 120°. A bent molecule with a bond angle of 120° has a central atom bonded to two atoms and one lone pair.
- A tetrahedral molecule has a central atom bonded to four atoms and no lone pairs, with bond angles of 109°. In a trigonal pyramidal molecule, a central atom is bonded to three atoms and one lone pair, with bond angles of 109°. In a bent molecule (109°), a central atom is bonded to two atoms and two lone pairs.
- The atoms bonded to the electron groups around the central atom determine the name of the three-dimensional geometry of a molecule.
- When the polar covalent bonds (dipoles) in a molecule have a symmetrical arrangement, the bond dipoles cancel each other; the molecule is nonpolar.
- When the polar covalent bonds (dipoles) in a molecule do not cancel each other, the molecule is polar.

Tutorial: Molecular Shape
Tutorial: Shapes of Molecules
Self Study Activity: The Shape of Molecules
Tutorial: Distinguishing Polar and Nonpolar Molecules

Guide to Predicting Molecular Shape (VSEPR Theory)	
Step 1	Draw the electron-dot formula.
Step 2	Arrange the electron groups around the central atom to minimize repulsion.
Step 3	Use the atoms bonded to the central atom to determine the molecular shape.

♦ **Learning Exercise 4.7A**

Use the number of electron groups and the number of bonded atoms to identify the electron group arrangement around a central atom and predict the shape of a molecule.

   a. linear   **b.** trigonal planar   **c.** tetrahedral

   d. trigonal pyramidal   **e.** bent (120°)   **f.** bent (109°)

	Electron Group Arrangement	Shape of Molecule
1. three electron groups with three bonded atoms	_____	_____
2. two electron groups with two bonded atoms	_____	_____
3. four electron groups with three bonded atoms	_____	_____
4. three electron groups with two bonded atoms	_____	_____
5. four electron groups with four bonded atoms	_____	_____
6. four electron groups with two bonded atoms	_____	_____

Answers   **1.** b, b   **2.** a, a   **3.** c, d   **4.** b, e   **5.** c, c   **6.** c, f

♦ **Learning Exercise 4.7B**

For each of the following, draw the electron-dot formula, state the number of electron groups and bonded atoms, and predict the shape and angle of the molecule.

Molecule	Electron-Dot Formula	Number of Electron Groups	Number of Bonded Atoms	Shape and Angle
CH_4				
PCl_3				
SO_3				
H_2S				

Answers

Molecule	Electron-Dot Formula	Number of Electron Groups	Number of Bonded Atoms	Shape and Angle
CH_4	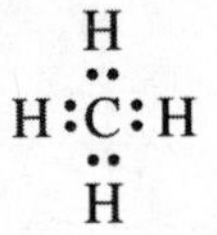 H H:C:H H	4	4	Tetrahedral, 109°
PCl_3	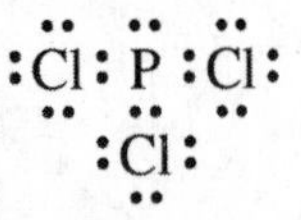 :Cl: P :Cl: :Cl:	4	3	Trigonal pyramidal, 109°
SO_3	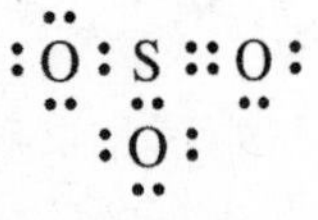 :O: S ::O: :O:	3	3	Trigonal planar, 120°
H_2S	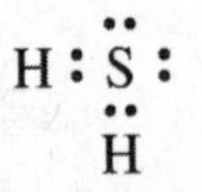 H: S : H	4	2	Bent, 109°

♦ **Learning Exercise 4.7C**

Determine whether each of the following is a polar molecule or a nonpolar molecule:

1. CF_4 2. HCl

3. NH_3 4. OF_2

Answers

1. In the tetrahedral CF_4 molecule, all C—F dipoles cancel; CF_4 is a nonpolar molecule.

2. In HCl, there is a single H—Cl dipole, HCl is a polar molecule.

3. In the trigonal pyramidal NH_3 molecule, the N—H dipoles do not cancel; NH_3 is a polar molecule.

4. In the bent OF_2 molecule, the O—F dipoles do not cancel; OF_2 is a polar molecule.

4.8 Attractive Forces in Compounds

- The interactions between particles in compounds determine their melting points.
- In polar molecules, dipole–dipole attractions occur between the positive end of one molecule and the negative end of another.
- Hydrogen bonding, a type of dipole–dipole interaction, occurs between partially positive hydrogen atoms and the strongly electronegative atoms fluorine, oxygen, or nitrogen.
- Dispersion forces occur when temporary dipoles form within nonpolar molecules, causing momentary weak attractions to temporary dipoles in other nonpolar molecules.
- Ionic solids have high melting points due to strong attractions between positive and negative ions.

MC

Tutorial: Intermolecular Forces

♦ **Learning Exercise 4.8**

Indicate the major type of interactive force that occurs in each of the following substances:

a. ionic bonds   **b.** dipole–dipole attractions   **c.** hydrogen bonds   **d.** dispersion forces

1. _____ KCl   2. _____ NCl_3 3. _____ SCl_2 4. _____ Cl_2
5. _____ HF   6. _____ H_2O 7. _____ C_4H_{10} 8. _____ Na_2O

Answers   1. a   2. b   3. b   4. d
              5. c   6. c   7. d   8. a

Checklist for Chapter 4

You are ready to take the Practice Test for Chapter 4. Be sure that you have accomplished the following learning goals for this chapter. If you are not sure, review the section listed at the end of the goal. Then apply your new skills and understanding to the Practice Test.

After studying Chapter 4, I can successfully:

_____ Illustrate the octet rule for the formation of ions (4.1).

_____ Write the formulas of compounds containing the ions of metals and nonmetals (4.2).

_____ Write the name of an ionic compound (4.3).

_____ Write the formula of a compound containing a polyatomic ion (4.4).

_____ Draw the electron-dot formula for a covalent compound (4.5).

_____ Write the name and formula of a covalent compound (4.5).

_____ Classify a bond as nonpolar covalent, polar covalent, or ionic (4.6).

_____ Use the VSEPR theory to determine the shape of a molecule (4.7).

_____ Classify a molecule as a polar or nonpolar molecule (4.7).

_____ Identify the attractive forces between particles (4.8).

Practice Test for Chapter 4

For questions 1 through 4, consider an atom of phosphorus.

1. It is in Group
   A. 2A (2)   **B.** 3A (13)   **C.** 5A (15)   **D.** 7A (17)   **E.** 8A (18)

2. How many valence electrons does it have?
   A. 2   **B.** 3   **C.** 5   **D.** 8   **E.** 15

3. To achieve an octet, the phosphorus atom will
   A. lose 1 electron   **B.** lose 2 electrons   **C.** lose 5 electrons
   D. gain 2 electrons   **E.** gain 3 electrons

4. As an ion, it has an ionic charge of
   A. 1+   **B.** 2+   **C.** 5+   **D.** 2–   **E.** 3–

5. To achieve an octet, a calcium atom
   A. loses 1 electron   **B.** loses 2 electrons   **C.** loses 3 electrons
   D. gains 1 electron   **E.** gains 2 electrons

6. To achieve an octet, a chlorine atom
   A. loses 1 electron   **B.** loses 2 electrons   **C.** loses 3 electrons
   D. gains 1 electron   **E.** gains 2 electrons

7. Another name for a positive ion is
   A. anion   **B.** cation   **C.** proton   **D.** positron   **E.** sodium

8. The correct ionic charge for calcium ion is
   A. 1+   **B.** 2+   **C.** 1–   **D.** 2–   **E.** 3–

9. The silver ion has a charge of
   A. 1+   **B.** 2+   **C.** 1–   **D.** 2–   **E.** 3–

10. The correct ionic charge for phosphate ion is
    A. 1+   **B.** 2+   **C.** 1–   **D.** 2–   **E.** 3–

11. The correct ionic charge for fluoride is
    A. 1+   **B.** 2+   **C.** 1–   **D.** 2–   **E.** 3–

12. The correct ionic charge for sulfate ion is
    A. 1+   **B.** 2+   **C.** 1–   **D.** 2–   **E.** 3–

13. When the elements magnesium and sulfur react,
    A. an ionic compound forms
    B. a nonpolar covalent compound forms
    C. no reaction occurs
    D. the two repel each other and won't combine
    E. none of the above

14. An ionic bond typically occurs between
    A. two different nonmetals
    B. two of the same type of nonmetals
    C. two noble gases
    D. two different metals
    E. a metal and a nonmetal

15. A nonpolar covalent bond typically occurs between
    A. two different nonmetals
    B. two of the same type of nonmetals
    C. two noble gases
    D. two different metals
    E. a metal and a nonmetal

16. A polar covalent bond typically occurs between
    A. two different nonmetals
    B. two of the same type of nonmetals
    C. two noble gases
    D. two different metals
    E. a metal and a nonmetal

17. The formula for a compound between carbon and chlorine is
    A. Cl   **B.** CCl_2 **C.** C_4Cl **D.** CCl_4 **E.** C_4Cl_2

18. The formula for a compound between sodium and sulfur is
    A. SoS   **B.** NaS   **C.** Na_2S **D.** NaS_2 **E.** Na_2SO_4

19. The formula for a compound between aluminum and oxygen is
    A. AlO   **B.** Al_2O **C.** AlO_3 **D.** Al_2O_3 **E.** Al_3O_2

20. The formula for a compound between barium and sulfur is
    A. BaS   **B.** BaS_2 **C.** BaS_2 **D.** Ba_2S_2 **E.** $BaSO_4$

21. The correct formula for iron(III) chloride is
    A. FeCl   **B.** $FeCl_2$ **C.** Fe_2Cl **D.** Fe_3Cl **E.** $FeCl_3$

22. The correct formula for ammonium sulfate is
  A. AmS **B.** $AmSO_4$ **C.** $(NH_4)_2S$ **D.** NH_4SO_4 **E.** $(NH_4)_2SO_4$

23. The correct formula for nickel(II) chloride is
  A. NCl **B.** NiCl **C.** Ni_2Cl_2 **D.** $NiCl_2$ **E.** Ni_2Cl

24. The correct formula for lithium phosphate is
  A. $LiPO_4$ **B.** Li_2PO_4 **C.** Li_3PO_4 **D.** $Li_2(PO_4)_3$ **E.** $Li_3(PO_4)_2$

25. The correct formula for silver oxide is
  A. AgO **B.** Ag_2O **C.** AgO_2 **D.** Ag_3O_2 **E.** Ag_3O

26. The correct formula for magnesium carbonate is
  A. $MgCO_3$ **B.** Mg_2CO_3 **C.** $Mg(CO_3)_2$ **D.** MgCO **E.** $Mg_2(CO_3)_3$

27. The correct formula for chromium(II) sulfate is
  A. $CrSO_3$ **B.** $CrSO_4$ **C.** Cr_2SO_3 **D.** $Cr(SO_4)_2$ **E.** Cr_2SO_4

28. The name of $AlPO_4$ is
  A. aluminum phosphide **B.** alum phosphate **C.** aluminum phosphate
  D. aluminum phosphorus oxide **E.** aluminum phosphite

29. The name of CuS is
  A. copper sulfide **B.** copper(I) sulfate **C.** copper(I) sulfide
  D. cuprous sulfide **E.** copper(II) sulfide

30. The name of $FeCl_2$ is
  A. ferric chloride **B.** iron(II) chlorine **C.** iron(II) chloride
  D. iron chlorine **E.** iron(III) chloride

31. The name of $ZnCO_3$ is
  A. zinc(III) carbonate **B.** zinc(II) carbonate **C.** zinc bicarbonate
  D. zinc carbon trioxide **E.** zinc carbonate

32. The name of Al_2O_3 is
  A. aluminum oxide **B.** aluminum(II) oxide **C.** aluminum trioxide
  D. dialuminum trioxide **E.** aluminum oxygenate

33. The name of NCl_3 is
  A. nitrogen chloride **B.** nitrogen trichloride **C.** trinitrogen chloride
  D. nitrogen chlorine three **E.** nitrogen chloride(III)

34. The name of CO is
  A. carbon monoxide **B.** carbonic oxide **C.** carbon oxide
  D. carbonious oxide **E.** carboxide

For questions 35 through 40, indicate the type of bonding expected between the following pairs of elements:

  A. ionic **B.** nonpolar covalent **C.** polar covalent **D.** none

35. ____ silicon and oxygen **36.** ____ chlorine and chlorine

37. ____ barium and chlorine   **38.** ____ sulfur and oxygen

39. ____ aluminum and fluorine   **40.** ____ neon and oxygen

For questions 41 through 46, determine the shape of each of the following molecules as

  A. linear   **B.** trigonal planar   **C.** tetrahedral
  D. trigonal pyramidal   **E.** bent (120°)   **F.** bent (109°)

41. has a central atom with three electron groups and three bonded atoms _____

42. has a central atom with two electron groups and two bonded atoms _____

43. PCl_3 _____

44. has a central atom with three electron groups and two bonded atoms _____

45. CCl_4 _____

46. H_2S _____

For questions 47 through 50, give the prevalent type of bonding

  A. ionic bonds   **B.** dipole–dipole attractions   **C.** hydrogen bonds   **D.** dispersion forces

47. NF_3

48. HF

49. LiCl

50. N_2

Answers to the Practice Test

1. C	**2.** C	**3.** E	**4.** E	**5.** B
6. D	**7.** B	**8.** B	**9.** A	**10.** E
11. C	**12.** D	**13.** A	**14.** E	**15.** B
16. A	**17.** D	**18.** C	**19.** D	**20.** A
21. E	**22.** E	**23.** D	**24.** C	**25.** B
26. A	**27.** B	**28.** C	**29.** E	**30.** C
31. E	**32.** A	**33.** B	**34.** A	**35.** C
36. B	**37.** A	**38.** C	**39.** A	**40.** D
41. B	**42.** A	**43.** D	**44.** E	**45.** C
46. F	**47.** B	**48.** C	**49.** A	**50.** D

Chemical Quantities and Reactions

Study Goals

- Use Avogadro's number to determine the number of particles in a given number of moles of an element or compound.
- Calculate the molar mass of a compound using its formula and the atomic masses on the periodic table.
- Use the molar mass to convert between the grams of a substance and the number of moles.
- Show that a balanced equation has an equal number of atoms of each element on the reactant side and the product side.
- Write a balanced equation for a chemical reaction when given the formulas of the reactants and products.
- Classify an equation as a combination, decomposition, single or double replacement, or combustion reaction.
- Describe the features of oxidation and reduction in an oxidation–reduction reaction.
- Using a given number of moles and a mole–mole conversion factor, determine the corresponding number of moles for a reactant or a product.
- Using molar masses and mole factors, convert between the grams of two substances in a reaction.
- Given the heat of reaction, describe the heat change in an exothermic reaction and in an endothermic reaction.
- Describe factors that affect the rate of a reaction.

Think About It

1. How is a mole analogous to a term for a collection such as one dozen?

2. How is a recipe like a chemical equation?

3. Why is the digestion of food a series of chemical reactions?

4. What are the different types of chemical reactions?

5. What causes a slice of apple or an avocado to turn brown?

6. In an exothermic reaction, is energy released or absorbed?

Key Terms

Match the following terms with the statements below:

 a. reactant   **b.** chemical equation   **c.** combination
 d. mole   **e.** molar mass   **f.** single replacement

1. ____ the amount of a substance that contains 6.02×10^{23} particles of that substance

2. ____ an initial substance that undergoes change in a chemical reaction

3. ____ the mass in grams of an element or compound that is equal numerically to its atomic mass or sum of atomic masses

4. ____ the type of reaction in which the reactants combine to form a single product

5. ____ a shorthand method of writing a chemical reaction with the formulas of the reactants written on the left side of an arrow and the formulas of the products on the right side

6. ____ the type of reaction in which an element replaces a different element in a reacting compound

Answers   **1.** d   **2.** a   **3.** e   **4.** c   **5.** b   **6.** f

5.1 The Mole

- A mole of any element contains Avogadro's number, 6.02×10^{23}, of atoms; a mole of any compound contains 6.02×10^{23} molecules or formula units.
- Avogadro's number provides a conversion factor between moles and number of particles:

$$\frac{1 \text{ mole}}{6.02 \times 10^{23} \text{ particles}} \quad \text{and} \quad \frac{6.02 \times 10^{23} \text{ particles}}{1 \text{ mole}}$$

- The subscripts in a formula indicate the number of moles of each element in one mole of the compound.

Guide to Calculating the Atoms or Molecules of a Substance	
Step 1	State the given and needed quantities.
Step 2	Write a plan to convert moles to atoms or molecules.
Step 3	Use Avogadro's number to write conversion factors.
Step 4	Set up the problem to calculate the number of particles.

Tutorial: Using Avogadro's Number

♦ **Learning Exercise 5.1A**

Calculate each of the following:

    a. number of P atoms in 1.50 moles of P

    b. number of H_2S molecules in 0.0750 mole of H_2S

    c. number of moles of Ag in 5.4×10^{24} atoms of Ag

    d. number of moles of C_3H_8 in 8.25×10^{24} molecules of C_3H_8

Answers   **a.** 9.03×10^{23} P atoms   **b.** 4.52×10^{22} H_2S molecules
                  c. 9.0 mole of Ag   **d.** 13.7 moles of C_3H_8

Study Note

The subscripts in the formula of a compound indicate the number of moles of each element in one mole of that compound. For example, consider the formula Mg_3N_2:

$$\text{1 mole of } Mg_3N_2 = \text{3 moles of Mg atoms}$$

$$\text{1 mole of } Mg_3N_2 = \text{2 moles of N atoms}$$

Conversion factors for the moles of elements in the above equalities can be written as

$$\frac{3 \text{ moles Mg atoms}}{1 \text{ mole } Mg_3N_2} = \frac{1 \text{ mole } Mg_3N_2}{3 \text{ moles Mg atoms}}$$

$$\frac{2 \text{ moles N atoms}}{1 \text{ mole } Mg_3N_2} = \frac{1 \text{ mole } Mg_3N_2}{2 \text{ moles N atoms}}$$

Guide to Calculating the Moles of an Element in a Compound	
Step 1	State the given and needed quantities.
Step 2	Write a plan to convert moles of a compound to moles of an element.
Step 3	Write equalities and conversion factors using subscripts.
Step 4	Set up the problem to calculate the moles of an element.

Tutorial: Moles and the Chemical Formula

♦ **Learning Exercise 5.1B**

Consider the formula for vitamin C (ascorbic acid), $C_6H_8O_6$.

    a. How many moles of carbon are in 2.0 moles of vitamin C?

    b. How many moles of hydrogen are in 5.0 moles of vitamin C?

    c. How many moles of oxygen are in 1.5 moles of vitamin C?

Answers   **a.** 12 moles of carbon (C)   **b.** 40. moles of hydrogen (H)
             c. 9.0 moles of oxygen (O)

♦ **Learning Exercise 5.1C**

For the compound ibuprofen ($C_{13}H_{18}O_2$) used in Advil and Motrin, calculate each of the following:

    a. moles of carbon (C) in 2.20 moles of ibuprofen

    b. moles of hydrogen (H) in 0.175 mole of ibuprofen

    c. moles of oxygen (O) in 0.750 mole of ibuprofen

    d. moles of ibuprofen that contain 36 moles of hydrogen (H)

Answers   **a.** 28.6 moles of C   **b.** 3.15 moles of H
             c. 1.50 moles of O   **d.** 2.0 moles of ibuprofen

5.2 Molar Mass

- The molar mass (g/mole) of an element is numerically equal to its atomic mass in grams.
- The molar mass (g/mole) of a compound is determined by multiplying the molar mass of each element by its subscript in the formula and adding the results.
- The two conversion factors for the molar mass of NaOH (40.0 g/mole) have the following form:

$$\frac{40.0 \text{ g NaOH}}{1 \text{ mole NaOH}} \quad \text{and} \quad \frac{1 \text{ mole NaOH}}{40.0 \text{ g NaOH}}$$

Guide to Calculating Molar Mass	
Step 1	Obtain the molar mass of each element.
Step 2	Multiply each molar mass by the number of moles (subscript) in the formula.
Step 3	Calculate the molar mass by adding the masses of the elements.

Example: What is the molar mass of silver nitrate, $AgNO_3$?

Step 1: Obtain the molar mass of each element.

$$Ag = 107.9 \text{ g}; \quad N = 14.0 \text{ g}; \quad O = 16.0 \text{ g}$$

Step 2: Multiply each molar mass by the number of moles in the formula.

$$1 \text{ mole Ag} \times \frac{107.9 \text{ g Ag}}{1 \text{ mole Ag}} = 107.9 \text{ g of Ag}$$

$$1 \text{ mole N} \times \frac{14.0 \text{ g N}}{1 \text{ mole N}} = 14.0 \text{ g of N}$$

$$3 \text{ moles O} \times \frac{16.0 \text{ g O}}{1 \text{ mole O}} = 48.0 \text{ g of O}$$

Step 3: Calculate the molar mass by adding the masses of the elements.

$$\text{Molar mass } AgNO_3 = (107.9 \text{ g} + 14.0 \text{ g} + 48.0 \text{ g}) = 169.9 \text{ g}$$

♦ **Learning Exercise 5.2A**

Determine the molar mass for each of the following:

a. K_2O

b. $AlCl_3$

c. $C_{13}H_{18}O_2$

d. C_4H_{10}

e. $Ca(NO_3)_2$

f. Mg_3N_2

g. $FeCO_3$

h. $(NH_4)_3PO_4$

Answers   **a.** 94.2 g   **b.** 133.5 g   **c.** 206 g   **d.** 58.1 g
                   e. 164.1 g   **f.** 100.9   **g.** 115.9 g   **h.** 149.1 g

	Guide to Calculating the Moles (or Grams) of a Substance from Grams (or Moles)
Step 1	State the given and needed quantities.
Step 2	Write a plan to convert moles to grams (or grams to moles).
Step 3	Determine the molar mass and write conversion factors.
Step 4	Set up the problem to convert moles to grams (or grams to moles).

Example: What is the mass in grams of 0.254 mole of Na_2CO_3 ?

Step 1: Given 0.254 mole of Na_2CO_3 **Need** mass in grams of Na_2CO_3
Step 2: Plan moles $\rightarrow$ grams
Step 3: Molar Mass and Conversion Factors

$$1 \text{ mole of } Na_2CO_3 \ = \ 106.0 \text{ g of } Na_2CO_3$$

$$\frac{106.0 \text{ g } Na_2CO_3}{1 \text{ mole } Na_2CO_3} \quad \text{and} \quad \frac{1 \text{ mole } Na_2CO_3}{106.0 \text{ g } Na_2CO_3}$$

Step 4: Set Up Problem

$$0.254 \text{ mole } Na_2CO_3 \ \times \frac{106.0 \text{ g } Na_2CO_3}{1 \text{ mole } Na_2CO_3} = 26.9 \text{ g of } Na_2CO_3$$

Molar Mass Factor

♦ **Learning Exercise 5.2B**

Calculate the number of grams in each of the following quantities:

a. 0.100 mole of SO_2

b. 0.100 mole of H_2SO_4

c. 2.50 moles of NH_3

d. 1.25 moles of O_2

e. 0.500 mole of Mg

f. 5.00 moles of H_2

g. 10.0 moles of PCl_3

h. 0.400 mole of S

Answers   **a.** 6.41 g   **b.** 9.81 g   **c.** 42.5 g   **d.** 40.0 g
        e. 12.2 g   **f.** 10.0 g   **g.** 1380 g   **h.** 12.8 g

---

Study Note

When the grams of a substance are given, the molar mass is used to calculate the number of moles of substance present.

$$\text{grams substance} \times \frac{1 \text{ mole substance}}{\text{grams substance}} = \text{mole of substance}$$

Example: How many moles of NaOH are in 4.00 g of NaOH?

$$4.00 \text{ g NaOH} \times \frac{1 \text{ mole NaOH}}{40.0 \text{ g NaOH}} = 0.100 \text{ mole of NaOH}$$

Molar Mass (inverted)

(MC)

Tutorial: Converting between Grams and Moles

Self Study Activity: Stoichiometry

♦ **Learning Exercise 5.2C**

Calculate the number of moles in each of the following:

a. 32.0 g of CH_4 **b.** 391 g of K

c. 8.00 g of C_3H_8 **d.** 25.0 g of Cl_2

e. 0.220 g of CO_2 **f.** 5.00 g of Al_2O_3

g. The methane used to heat a home has the formula CH_4. If 725 grams of methane are used, how many moles of methane were used during this time?

h. A vitamin tablet contains 18 mg of iron. If there are 100 tablets in a bottle, how many moles of iron are in the vitamins in the bottle?

Answers	**a.** 2.00 moles	**b.** 10.0 moles	**c.** 0.182 mole
	d. 0.352 mole	**e.** 0.00500 mole	**f.** 0.0490 mole
	g. 45.3 moles	**h.** 0.032 mole	

5.3 Chemical Reactions and Equations

- A chemical change occurs when the atoms of the initial substances rearrange to form new substances that have new properties.
- A chemical equation shows the formulas of the reactants on the left side of the arrow and the formulas of the products on the right side.
- Each formula in an equation is followed by an abbreviation, in parentheses, that gives the physical state of the substance: solid (*s*), liquid (*l*), or gas (*g*), or, if dissolved in water, an aqueous solution (*aq*).
- The Greek letter delta (Δ) over the arrow in an equation represents the application of heat to the reaction.
- In a balanced equation, *coefficients* in front of the formulas provide the same number of atoms for each kind of element on the reactant and product sides.

(MC)

Tutorial: Chemical Reactions and Equations

Self Study Activity: Chemical Reactions and Equations

◆ **Learning Exercise 5.3A**

State the number of atoms of each element on the reactant and the product side for each equation:

a. $CaCO_3(s) \rightarrow CaO(s) + CO_2(g)$

Element	Atoms on reactant side	Atoms on product side
Ca		
C		
O		

b. $2Na(s) + H_2O(l) \rightarrow Na_2O(s) + H_2(g)$

Element	Atoms on reactant side	Atoms on product side
Na		
H		
O		

c. $C_5H_{12}(l) + 8O_2(g) \rightarrow 5CO_2(g) + 6H_2O(g)$

Element	Atoms on reactant side	Atoms on product side
C		
H		
O		

d. $2AgNO_3(aq) + K_2S\ (aq) \rightarrow 2KNO_3(aq) + Ag_2S(s)$

Element	Atoms on reactant side	Atoms on product side
Ag		
N		
O		
K		
S		

e. $2Al(OH)_3(s) + 3H_2SO_4(aq) \rightarrow Al_2(SO_4)_3(aq) + 6H_2O(l)$

Element	Atoms on reactant side	Atoms on product side
Al		
O		
H		
S		

Answers

a. $CaCO_3(s) \rightarrow CaO(s) + CO_2(g)$

Element	Atoms on reactant side	Atoms on product side
Ca	1	1
C	1	1
O	3	3

b. $2Na(s) + H_2O(l) \rightarrow Na_2O(s) + H_2(g)$

Element	Atoms on reactant side	Atoms on product side
Na	2	2
H	2	2
O	1	1

c. $C_5H_{12}(l) + 8O_2(g) \rightarrow 5CO_2(g) + 6H_2O(g)$

Element	Atoms on reactant side	Atoms on product side
C	5	5
H	12	12
O	16	16

d. $2AgNO_3(aq) + K_2S(aq) \rightarrow 2KNO_3(aq) + Ag_2S(s)$

Element	Atoms on reactant side	Atoms on product side
Ag	2	2
N	2	2
O	6	6
K	2	2
S	1	1

e. $2Al(OH)_3(s) + 3H_2SO_4(aq) \rightarrow Al_2(SO_4)_3(aq) + 6H_2O(l)$

Element	Atoms on reactant side	Atoms on product side
Al	2	2
O	18	18
H	12	12
S	3	3

Tutorial: Balancing Chemical Equations

Guide to Balancing a Chemical Equation	
Step 1	Write an equation using the correct formulas of the reactants and products.
Step 2	Count the atoms of each element in reactants and products.
Step 3	Use coefficients to balance each element.
Step 4	Check the final equation for balance.

♦ **Learning Exercise 5.3B**

Balance each of the following by placing appropriate coefficients in front of the formulas as needed:

a. ____ $MgO(s) \rightarrow$ ____ $Mg(s) +$ ____ $O_2(g)$

b. ____ $Zn(s) +$ ____ $HCl(aq) \rightarrow$ ____ $ZnCl_2(aq) +$ ____ $H_2(g)$

c. ____ $Al(s) +$ ____ $CuSO_4(aq) \rightarrow$ ____ $Cu(s) +$ ____ $Al_2(SO_4)_3(aq)$

d. ____ $Al_2S_3(s) +$ ____ $H_2O(l) \rightarrow$ ____ $Al(OH)_3(s) +$ ____ $H_2S(g)$

e. ____ $BaCl_2(aq) +$ ____ $Na_2SO_4(aq) \rightarrow$ ____ $BaSO_4(s) +$ ____ $NaCl(aq)$

f. ____ $CO(g) +$ ____ $Fe_2O_3(s) \rightarrow$ ____ $Fe(s) +$ ____ $CO_2(g)$

g. ____ $K(s) +$ ____ $H_2O(l) \rightarrow$ ____ $K_2O(s) +$ ____ $H_2(g)$

h. ____ $Fe(OH)_3(s) \xrightarrow{\Delta}$ ____ $Fe_2O_3(s) +$ ____ $H_2O(l)$

Answers

a. $2MgO(s) \rightarrow 2Mg(s) + O_2(g)$

b. $Zn(s) + 2HCl(aq) \rightarrow ZnCl_2(aq) + H_2(g)$

c. $2Al(s) + 3CuSO_4(aq) \rightarrow 3Cu(s) + Al_2(SO_4)_3(aq)$

d. $Al_2S_3(s) + 6H_2O(l) \rightarrow 2Al(OH)_3(s) + 3H_2S(g)$

e. $BaCl_2(aq) + Na_2SO_4(aq) \rightarrow BaSO_4(s) + 2NaCl(aq)$

f. $3CO(g) + Fe_2O_3(s) \rightarrow 2Fe(s) + 3CO_2(g)$

g. $2K(s) + H_2O(l) \rightarrow K_2O(s) + H_2(g)$

h. $2Fe(OH)_3(s) \xrightarrow{\Delta} Fe_2O_3(s) + 3H_2O(l)$

5.4 Types of Reactions

- Reactions are classified as combination, decomposition, single replacement, double replacement, and combustion.
- In a *combination* reaction, reactants are combined. In a *decomposition* reaction, a reactant splits into simpler products.
- In a *single replacement* reaction, one element in a reacting compound is replaced with another element. In a *double replacement* reaction, parts of two different reactants exchange places.
- In a *combustion* reaction, a compound of carbon and hydrogen reacts with oxygen to form CO_2, H_2O, and energy.

◆ **Learning Exercise 5.4A**

Match each of the following reactions with the type of reaction:

    a. combination   **b.** decomposition

    c. single replacement   **d.** double replacement

    e. combustion

1. ____ $N_2(g) + 3H_2(g) \rightarrow 2\,NH_3(g)$

2. ____ $BaCl_2(aq) + K_2CO_3(aq) \rightarrow BaCO_3(s) + 2\,KCl(aq)$

3. ____ $C_3H_8(g) + 5O_2(g) \xrightarrow{\Delta} 3CO_2(g) + 4O_2(g)$

4. ____ $CuO(s) + H_2(g) \rightarrow Cu(s) + H_2O(l)$

5. ____ $N_2(g) + 2O_2(g) \rightarrow 2NO_2(g)$

6. ____ $2NaHCO_3(s) \xrightarrow{\Delta} Na_2O(s) + 2CO_2(g) + H_2O(l)$

7. ____ $PbCO_3(s) \xrightarrow{\Delta} PbO(s) + CO_2(g)$

8. ____ $Al(s) + Fe_2O_3(s) \rightarrow Fe(s) + Al_2O_3(s)$

Answers   **1.** a   **2.** d   **3.** e   **4.** c

                 5. a   **6.** b   **7.** b   **8.** c

◆ **Learning Exercise 5.4B**

1. One way to remove tarnish from silver is to place the silver object on a piece of aluminum foil and add boiling water and some baking soda. The unbalanced equation is the following:

    $Al(s) + Ag_2S(s) \rightarrow Ag(s) + Al_2S_3(s)$

    a. What is the balanced equation?

    b. What type of reaction takes place?

2. Octane, C_8H_{18}, a compound in gasoline, burns in oxygen to produce carbon dioxide and water. All compounds are gases.

    a. What is the balanced equation for the reaction?

    b. What type of reaction takes place?

Answers   **1a.** $2Al(s) + 3Ag_2S(s) \rightarrow 6Ag(s) + Al_2S_3(s)$ **1b.** single replacement

               2a. $2C_8H_{18}(g) + 25O_2(g) \xrightarrow{\Delta} 16CO_2(g) + 18H_2O(g)$ **2b.** combustion

5.5 Oxidation–Reduction Reactions

- In an oxidation–reduction reaction, there is a loss and gain of electrons. In an oxidation, electrons are lost. In a reduction, electrons are gained.
- An oxidation is always accompanied by a reduction. The number of electrons lost in the oxidation reaction is equal to number of electrons gained in the reduction reaction.
- In biological systems, the term *oxidation* describes the gain of oxygen or the loss of hydrogen. The term *reduction* is used to describe a loss of oxygen or a gain of hydrogen.

Tutorial: Identifying Oxidation–Reduction Reactions

♦ **Learning Exercise 5.5**

For each of the following reactions, indicate whether the underlined element is *oxidized* or *reduced*.

    a. $4\underline{Al}(s) + 3O_2(g) \rightarrow 2Al_2O_3(s)$ Al is _____

    b. $\underline{Fe}^{3+}(aq) + 1\ e^- \rightarrow Fe^{2+}(aq)$ Fe^{3+} is _____

    c. $\underline{Cu}O(s) + H_2(g) \rightarrow Cu(s) + H_2O(l)$ Cu^{2+} is _____

    d. $2\underline{Cl}^-(aq) \rightarrow Cl_2(g) + 2\ e^-$ Cl^- is _____

    e. $2H\underline{Br}(aq) + Cl_2(g) \rightarrow 2HCl(aq) + Br_2(g)$ Br^- is _____

    f. $2\underline{Na}(s) + Cl_2(g) \rightarrow 2NaCl(s)$ Na is _____

    g. $\underline{Cu}Cl_2(aq) + Zn(g) \rightarrow ZnCl_2(aq) + Cu(g)$ Cu^{2+} is _____

Answers

    a. Al is oxidized to Al^{3+}; loss of electrons

    b. Fe^{3+} is reduced to Fe^{2+}; gain of electrons

    c. Cu^{2+} is reduced to Cu; gain of electrons

    d. Cl^- is oxidized to Cl_2; loss of electrons

    e. Br^- is oxidized to Br_2; loss of electrons

    f. Na is oxidized to Na^+; loss of electrons

    g. Cu^{2+} is reduced to Cu; gain of electrons

5.6 Mole Relationships in Chemical Equations

- The law of conservation of mass states that mass is not lost or gained during a chemical reaction.
- The coefficients in a balanced chemical equation describe the moles of reactants and products in the reactions.
- Using the coefficients, mole–mole conversion factors are written for any two substances in the equation.
- For the reaction of oxygen forming ozone, $3O_2(g) \rightarrow 2O_3(g)$, the mole–mole conversion factors are the following:

$$\frac{3 \text{ moles } O_2}{2 \text{ moles } O_3} \quad \text{and} \quad \frac{2 \text{ moles } O_3}{3 \text{ moles } O_2}$$

Self Study Activity: Stoichiometry
Tutorial: Law of Conservation of Mass
Tutorial: Moles of Reactants and Products
Tutorial: Conversions Involving Moles

♦ **Learning Exercise 5.6A**

Write the conversion factors that are possible from the following equation: $N_2(g) + O_2(g) \rightarrow 2NO(g)$

Answers

$$\frac{1 \text{ mole } N_2}{1 \text{ mole } O_2} \quad \text{and} \quad \frac{1 \text{ mole } O_2}{1 \text{ mole } N_2}; \quad \frac{1 \text{ mole } N_2}{2 \text{ moles } NO} \quad \text{and} \quad \frac{2 \text{ moles } NO}{1 \text{ mole } N_2}$$

$$\frac{1 \text{ mole } O_2}{2 \text{ moles } NO} \quad \text{and} \quad \frac{2 \text{ moles } NO}{1 \text{ mole } O_2}$$

Guide to Using Mole–Mole Factors	
Step 1	State the given and needed quantities.
Step 2	Write a plan to convert the given to the needed moles.
Step 3	Use coefficients to write relationships and mole–mole factors.
Step 4	Set up the problem using the mole–mole factor that cancels given moles.

> ## Study Note
>
> The appropriate mole–mole factor is used to change the number of moles of the given to moles of the needed.
>
> *Example:* Using the equation, $N_2(g) + O_2(g) \rightarrow 2NO(g)$, calculate the moles of NO obtained from 3.0 moles of N_2.
>
> **Step 1: Given** 3.0 moles of N_2 **Need** moles of NO
>
> **Step 2: Plan** moles of $N_2 \rightarrow$ moles of NO
>
> **Step 3: Mole–Mole Factor** $\dfrac{2 \text{ moles NO}}{1 \text{ mole } N_2}$
>
> **Step 4: Set Up Problem** $3.0 \text{ moles } N_2 \times \dfrac{2 \text{ moles NO}}{1 \text{ mole } N_2} = 6.0 \text{ moles of NO}$

♦ **Learning Exercise 5.6B**

Use the equation below to answer the following questions:

$$C_3H_8(g) + 5O_2(g) \rightarrow 3CO_2(g) + 4H_2O(g)$$

a. How many moles of O_2 are needed to react with 2.00 moles of C_3H_8?

b. How many moles CO_2 are produced when 4.00 moles of O_2 react?

c. How many moles of C_3H_8 react with 3.00 moles of O_2?

d. How many moles of H_2O are produced from 0.50 mole of C_3H_8?

Answers   **a.** 10.0 moles of O_2 **b.** 2.40 moles of CO_2
               c. 0.600 mole of C_3H_8 **d.** 2.0 moles of H_2O

5.7 Mass Calculations for Reactions

- The grams or moles of a substance in an equation are converted to another using molar masses and mole–mole factors.

Guide to Calculating the Masses of Reactants and Products in a Chemical Reaction	
Step 1	Use molar mass to convert grams of given to moles (if necessary).
Step 2	Write a mole–mole factor from the coefficients in the equation.
Step 3	Convert moles of given to moles of needed substance using the mole–mole factor.
Step 4	Convert moles of needed substance to grams using the molar mass.

Tutorial: Masses of Reactants and Products
Self Study Activity: Stoichiometry

Study Note

How many grams of NO can be produced from 8.00 g of O_2?

Equation: $N_2(g) + O_2(g) \rightarrow 2NO(g)$

Solution: **Step 1**   **Step 2 and 3**   **Step 4**

$$8.00 \text{ g } O_2 \times \frac{1 \text{ mole } O_2}{32.0 \text{ g } O_2} \times \frac{2 \text{ moles NO}}{1 \text{ mole } O_2} \times \frac{30.0 \text{ g NO}}{1 \text{ mole NO}} = 15.0 \text{ g of NO}$$

◆ **Learning Exercise 5.7**

Consider the equation for the following questions:

$$2C_2H_6(g) + 7O_2(g) \rightarrow 4CO_2(g) + 6H_2O(g)$$

a. How many grams of oxygen (O_2) are needed to react with 4.00 moles of C_2H_6?

b. How many grams of C_2H_6 are needed to react with 115 g of O_2?

c. How many grams of C_2H_6 react if 22.0 g of CO_2 gas is produced?

d. How many grams of CO_2 are produced when 2.00 moles of C_2H_6 react with sufficient oxygen?

e. How many grams of water are produced when 82.5 g of O_2 react with sufficient C_2H_6?

Answers   **a.** 448 g of O_2 **b.** 30.8 g of C_2H_6 **c.** 7.53 g of C_2H_6
             d. 176 g of CO_2 **e.** 39.8 g of H_2O

5.8 Energy in Chemical Reactions

- In a reaction, molecules (or atoms) must collide with energy equal to or greater than the energy of activation.
- The heat of reaction is the energy difference between the energy of the reactants and the products.
- In exothermic reactions, the heat of reaction is the energy released. In endothermic reactions, the heat of reaction is the energy absorbed.
- Generally, the rate of a reaction (the speed at which products form) can be increased by adding more reacting molecules, raising the temperature of the reaction, or by adding a catalyst.

Tutorial: Heat of Reaction
Tutorial: Activation Energy and Transition State
Tutorial: Factors that Affect Rate

◆ **Learning Exercise 5.8A**

Indicate whether each of the following is an endothermic or exothermic reaction:

1. $2H_2(g) + O_2(g) \rightarrow 2H_2O(g) + 582 \text{ kJ}$ _____

2. $C_2H_4(g) + 176 \text{ kJ} \rightarrow H_2(g) + C_2H_2(g)$ _____

3. $2C(s) + O_2(g) \rightarrow 2CO(g) + 220 \text{ kJ}$ _____

4. $C_6H_{12}O_6(s) + 6 O_2(g) \rightarrow 6CO_2(g) + 6 H_2O(l) + 1350 \text{ kcal}$ _____
   Glucose

5. $C_2H_4(g) + H_2O(g) \rightarrow C_2H_5OH(l) + 21 \text{ kcal}$ _____

Answers: **1.** exothermic   **2.** endothermic   **3.** exothermic
              4. exothermic   **5.** exothermic

◆ **Learning Exercise 5.8B**

Indicate how each of the following will affect the rate of the reaction for the combustion of propane:

$$C_3H_8(g) + 5O_2(g) \rightarrow 3CO_2(g) + 4H_2O(g)$$

a. adding propane

b. adding a catalyst

c. removing some oxygen

d. decreasing the temperature

Answers::   **a.** increase   **b.** increase   **c.** decrease
               d. decrease

Checklist for Chapter 5

You are ready to take the practice test for Chapter 5. Be sure that you have accomplished the following learning goals for this chapter. If you are not sure, review the section listed at the end of the goal. Then apply your new skills and understanding to the practice test.

After studying Chapter 5, I can successfully:

_____ Calculate the number of particles in a mole of a substance (5.1).

_____ Calculate the molar mass given the formula of a substance (5.2).

_____ Convert the grams of a substance to moles; moles to grams (5.2).

_____ Write a balanced equation for a chemical reaction from the formulas of the reactants and products (5.3).

_____ Identify a reaction as a combination, decomposition or single or double replacement or combustion (5.4).

_____ Identify an oxidation or reduction reaction (5.5).

_____ Use mole–mole factors from the mole relationships in an equation to calculate the moles of another substance in an equation for a chemical reaction (5.6).

_____ Calculate the mass of a substance in an equation using mole factors and molar masses (5.7).

_____ Identify exothermic and endothermic reactions from the heat in a chemical reaction (5.8).

_____ Determine how a change in conditions will affect the rate of a reaction (5.8).

Practice Test for Chapter 5

For questions 1 through 4, calculate the moles of each of the elements in the formula.

Tagamet has the formula $C_{10}H_{16}N_6S$. Tagamet is used to inhibit the production of acid in the stomach. Select the correct answer for the number of moles of each of the following in 0.25 mole of Tagamet.

    A. 0.25   **B.** 0.50   **C.** 1.5   **D.** 2.5   **E.** 4.0

1. moles of C   **2.** moles of H

3. moles of N   **4.** moles of S

5. The atoms of Na in 0.0500 moles of Na is
    A. 0.500 atoms   **B.** 6.02×10^{23} atoms   **C.** 3.01×10^{23} atoms
    D. 3.01×10^{22} atoms   **E.** 3.01×10^{25} atoms

6. What is the molar mass of K_2SO_4?
    A. 55.1 g   **B.** 87.2 g   **C.** 126.3 g   **D.** 135.2 g   **E.** 174.3 g

7. What is the molar mass of $NaNO_3$?
    A. 34.0 g   **B.** 37.0 g   **C.** 53.0 g   **D.** 75.0 g   **E.** 85.0 g

8. The number of grams in 0.600 mole of Cl_2 is
    A. 71.0 g   **B.** 118 g   **C.** 42.6 g   **D.** 84.5 g   **E.** 4.30 g

9. How many grams are in 4.00 moles of NH_3?
    A. 4.00 g   **B.** 17.0 g   **C.** 34.0 g   **D.** 68.0 g   **E.** 0.240 g

10. How many moles is 8.0 g of NaOH?
    A. 0.10 mole   **B.** 0.20 mole   **C.** 0.40 mole   **D.** 2.0 moles   **E.** 4.0 moles

11. The number of moles of aluminum in 54 g of Al is
    A. 0.20 mole   **B.** 1.0 mole   **C.** 2.0 moles   **D.** 3.0 moles   **E.** 4.0 moles

12. The number of moles of water in 3.6 g of H_2O is
    A. 0.20 mole   **B.** 1.0 mole   **C.** 2.0 moles   **D.** 3.0 moles   **E.** 4.0 moles

13. What is the number of moles in 2.2 g of CO_2?
    A. 2.0 moles   **B.** 1.0 mole   **C.** 0.20 mole   **D.** 0.050 mole   **E.** 0.010 mole

14. 0.20 g H_2 = ____ H_2
    A. 0.10 mole   **B.** 0.20 mole   **C.** 0.40 mole   **D.** 0.040 mole   **E.** 0.010 mole

For questions 15 through 19, indicate the correct coefficient for the component in the balanced equation written in boldface type.

    A. 1   **B.** 2   **C.** 3   **D.** 4   **E.** 5

15. ____ $Sn(s) + \mathbf{Cl_2}(g) \rightarrow SnCl_4(s)$

16. ____ $Al(s) + H_2O(l) \rightarrow Al_2O_3(s) + \mathbf{H_2}(g)$

17. ____ $C_3H_8(g) + \mathbf{O_2}(g) \rightarrow CO_2(g) + H_2O(g)$

18. ____ $\mathbf{NH_3}(g) + O_2(g) \rightarrow N_2(g) + H_2O(g)$

19. ____ $N_2O(g) \rightarrow N_2(g) + \mathbf{O_2}(g)$

For questions 20 through 24, classify each reaction as one of the following:

    A. combination   **B.** decomposition   **C.** single replacement
    D. double replacement   **E.** combustion

20. ____ $S(s) + O_2(g) \rightarrow SO_2(g)$

21. ____ $Fe_2O_3(s) + 3C(s) \rightarrow 2Fe(s) + 3CO(g)$

22. ____ $CaCO_3(s) \rightarrow CaO(s) + CO_2(g)$

23. ____ $Mg(s) + 2AgNO_3(aq) \rightarrow Mg(NO_3)_2(aq) + 2Ag(s)$

24. ____ $C_6H_{12}(g) + 9O_2(g) \rightarrow 6CO_2(g) + 6H_2O(g)$

25. ____ $Na_2S(aq) + Pb(NO_3)_2(aq) \rightarrow PbS(s) + 2NaNO_3(aq)$

For questions 26 through 30, identify as an (O) oxidation or a (R) reduction.

26. $Ca \rightarrow Ca^{2+} + 2e^-$ ____   **27.** $Fe^{3+} + 3e^- \rightarrow Fe$ ____

28. $Al^{3+} + 3e^- \rightarrow Al$ ____   **29.** $Br_2 + 2e^- \rightarrow 2Br^-$ ____

30. $Sn^{2+} \rightarrow Sn^{4+} + 2e^-$ ____

For questions 31 through 35, use the reaction:

$$C_2H_5OH(l) + 3O_2(g) \rightarrow 2CO_2(g) + 3H_2O(g)$$
Ethanol

31. How many grams of oxygen are needed to react with 1.0 mole of ethanol?
    A. 8.0 g   **B.** 16 g   **C.** 32 g   **D.** 64 g   **E.** 96 g

32. How many moles of water are produced when 12 moles of oxygen react?
    A. 3.0 moles   **B.** 6.0 moles   **C.** 8.0 moles   **D.** 12.0 moles   **E.** 36.0 moles

33. How many grams of carbon dioxide are produced when 92 g of ethanol react?
    A. 22 g   **B.** 44 g   **C.** 88 g   **D.** 92 g   **E.** 176 g

34. How many moles of oxygen would be needed to produce 44 g of CO_2?
    A. 0.67 mole   **B.** 1.0 mole   **C.** 1.5 moles   **D.** 2.0 moles   **E.** 3.0 moles

35. How many grams of water will be produced if 23 g of ethanol react?
    A. 54 g   **B.** 27 g   **C.** 18 g   **D.** 9.0 g   **E.** 6.0 g

For questions 36 through 38, indicate if the reactions are (EX) exothermic or (EN) endothermic:

36. $N_2(g) + 3H_2(g) \rightarrow 2NH_3(g) + 22$ kcal _____

37. $2HCl(g) + 44$ kcal $\rightarrow H_2(g) + Cl_2(g)$ _____

38. $C_3H_8(g) + 5O_2(g) \rightarrow 3CO_2(g) + 4H_2O(g) + 531$ kcal _____

For questions 39 and 40, indicate the effect of each of the following changes on the reaction:

$$C_2H_5OH(l) + 3O_2(g) \rightarrow 2CO_2(g) + 3H_2O(g)$$
Ethanol

I Increase the rate of the reaction   *D decrease the rate of the reaction*

39. adding O_2 _____

40. removing ethanol _____

Answers to the Practice Test

1. D	2. E	3. C	4. A	5. D
6. E	7. E	8. C	9. D	10. B
11. C	12. A	13. D	14. A	15. B
16. C	17. E	18. D	19. A	20. A
21. C	22. B	23. C	24. E	25. D
26. O	27. R	28. R	29. R	30. O
31. E	32. D	33. E	34. C	35. B
36. EX	37. EN	38. EX	39. I	40. D

Study Goals

- Use the kinetic molecular theory of gases to describe the properties of gases.
- Describe the units of measurement used for pressure, and change from one unit to another.
- Use the pressure–volume relationship (Boyle's law) to determine the new pressure or volume of a certain amount of gas at a constant temperature.
- Use the volume–temperature relationship (Charles's law) to determine the new temperature or volume of a certain amount of gas at a constant pressure.
- Use the pressure–temperature relationship (Gay-Lussac's law) to determine the new temperature or pressure of a certain amount of gas at a constant volume.
- Use the combined gas law to find the new pressure, volume, or temperature of a gas when changes in two of these properties are given.
- Describe the relationship between the amount of a gas and its volume (Avogadro's law), and use this relationship in calculations.
- Use partial pressures to calculate the total pressure of a mixture of gases.
- Determine the mass or volume of a gas at STP.

Think About It

1. How does a barometer work?

2. Why must the pressure of the inhaled gas mixture increase when a person is scuba diving?

3. Why are airplanes pressurized?

4. How do we picture a gas on the molecular level?

Key Terms

Match each of the following key terms with the correct definition:

- **a.** kinetic molecular theory   **b.** pressure   **c.** Boyle's law
- **d.** Charles's law   **e.** STP   **f.** partial pressure

1. _____ the volume of a gas varies directly with the Kelvin temperature when pressure and amount of gas remain constant

2. _____ a force exerted by gas particles that collide with the sides of a container

3. _____ the pressure exerted by the individual gases in a gas mixture

4. _____ the volume of a gas varies inversely with the pressure of a gas when temperature and amount of gas are constant

5. _____ a model that explains the behavior of gaseous particles

6. _____ condition of a gas at 1.00 atm and 0 °C

Answers   **1.** d   **2.** b   **3.** f   **4.** c   **5.** a   **6.** e

6.1 Properties of Gases

- In a gas, particles are far apart and moving so fast that any attraction between them would be minimal.
- A gas is described by the physical properties of pressure (P), volume (V), temperature (T), and amount in moles (n).

(MC)

Self Study Activity: Properties of Gases
Tutorial: Kinetic Molecular Theory of Gases

♦ **Learning Exercise 6.1**

Answer true (T) or false (F) for each of the following:

a _____ Gases are composed of small particles.

b. _____ Gas molecules are usually close together.

c. _____ Gas molecules move slowly because they are strongly attracted.

d. _____ Distances between gas molecules are large.

e. _____ Gas molecules travel in straight lines until they collide.

f. _____ Pressure is the force of gas particles striking the walls of a container.

g. _____ Kinetic energy decreases with increasing temperature.

Answers   **a.** T   **b.** F   **c.** F   **d.** T
                 e. T   **f.** T   **g.** F

6.2 Gas Pressure

- A gas exerts pressure, which is the force that gas particles exert on the walls of a container.
- Units of gas pressure include torr, mmHg, Pascal (Pa), kilopascal (kPa), lb/in.^{2} (psi), and atm.
- Gas pressure can be converted from one unit to another using conversion factors.

Tutorial: Converting between Units of Pressure
Case Study: Scuba Diving and Blood Gases

◆ **Learning Exercise 6.2**

Complete the following:

a. 1.50 atm = _____ mmHg

b. 150 mmHg = _____ atm

c. 0.725 atm = _____ kPa

d. 1520 mmHg = _____ atm

e. 30.5 psi = _____ mmHg

f. During the weather report on TV, the pressure is given as 29.4 in. (of mercury). What is this pressure in mmHg and in atm?

Answers   **a.** 1140 mmHg   **b.** 0.20 atm   **c.** 73.5 kPa
              d. 2.00 atm   **e.** 1580 mmHg   **f.** 747 mmHg; 0.983 atm

6.3 Pressure and Volume (Boyle's Law)

- According to Boyle's law, pressure increases if volume decreases and pressure decreases if volume increases.
- If two properties change in opposite directions, the properties have an *inverse relationship*.
- The volume (V) of a gas changes inversely with the pressure (P) of the gas when T and n are held constant: $P_1V_1 = P_2V_2$. The subscripts 1 and 2 represent initial and final conditions.

Guide to Using the Gas Laws	
Step 1	Organize the data in a table of initial and final conditions.
Step 2	Rearrange the gas law equation to solve for the unknown quantity.
Step 3	Substitute values into the gas law equation and calculate.

Tutorial: Pressure and Volume

♦ **Learning Exercise 6.3A**

Complete each of the following with increases or decreases:

1. Gas pressure increases (T and n constant) when volume _____.

2. Gas volume increases (T and n constant) when pressure _____.

Answers   1. decreases   2. decreases

♦ **Learning Exercise 6.3B**

Calculate the unknown quantity in each of the following gas problems using Boyle's law:

a. A sample of 4.0 L of helium gas has a pressure of 800. mmHg. What is the new pressure, in mmHg, when the volume is reduced to 1.0 L (n and T constant)?

b. A gas occupies a volume of 360 mL at 750 mmHg. What volume, in milliliters, does it occupy at a pressure of 375 mmHg (n and T constant)?

c. A gas sample at a pressure of 5.00 atm has a volume of 3.00 L. If the gas pressure is changed to 760. mmHg, what is the new volume, in liters, of the gas (n and T constant)?

d. A sample of 250. mL of nitrogen is initially at a pressure of 2.50 atm. If the pressure changes to 825 mmHg, what is the new volume, in liters (n and T constant)?

Answers   a. 3200 mmHg   b. 720 mL
                   c. 15.0 L   d. 0.576 L

6.4 Temperature and Volume (Charles's Law)

- According to Charles's law, volume increases if the temperature of the gas increases and volume decreases if temperature decreases.
- In a direct relationship, two properties increase or decrease together.
- The volume (V) of a gas is directly related to its Kelvin temperature (T) when there is no change in the pressure or moles of the gas:

$$\frac{V_1}{T_1} = \frac{V_2}{T_2}$$

MC

Tutorial: Temperature and Volume

♦ **Learning Exercise 6.4A**

Complete each of the following with increases or decreases:

a. When the temperature of a gas increases at constant pressure and amount, its volume _____.

b. When the volume of a gas decreases at constant pressure and amount, its temperature _____.

Answers   a. increases   b. decreases

♦ **Learning Exercise 6.4B**

Solve each of the following gas law problems using Charles's law:

a. A balloon has a volume of 2.5 L at a temperature of 0 °C. What is the new volume of the balloon when the temperature rises to 120 °C and the pressure and number of moles remain constant?

b. Consider a balloon filled with helium to a volume of 6600 L at a temperature of 223 °C. To what temperature, in degrees Celsius, must the gas be cooled to decrease the volume to 4100 L (*P* and *n* constant)?

c. A sample of 750 mL of neon is heated from 120 °C to 350 °C. If pressure and amount remain constant, what is the new volume, in milliliters?

d. What is the final temperature, in degrees Celsius, of 350 mL of oxygen gas at 22 °C when its volume decreases to 0.100 L (*P* and *n* constant)?

Answers   a. 3.6 L   b. 35 °C   c. 1200 mL   d. −189 °C

6.5 Temperature and Pressure (Gay-Lussac's Law)

- According to Gay-Lussac's law, at constant volume and amount, the gas pressure increase when the temperature of the gas increases, and the pressure decreases when the temperature decreases.
- The pressure (*P*) of a gas is directly related to its Kelvin temperature (*T*), when *n* and *V* are constant:

$$\frac{P_1}{T_1} = \frac{P_2}{T_2}$$ The subscripts 1 and 2 represent initial and final conditions.

Vapor pressure is the pressure of the gas that forms when a liquid evaporates. At the boiling point of a liquid, the vapor pressure equals the atmospheric pressure.

Tutorial: Temperature and Pressure
Tutorial: Vapor Pressure and Boiling Point

◆ **Learning Exercise 6.5A**

Solve the following gas law problems using Gay-Lussac's law:

   a. A sample of helium gas has a pressure of 860. mmHg at a temperature of 225 K. At what pressure (mmHg) will the helium sample reach a temperature of 675 K (V and n constant)?

   b. A balloon contains a gas with a pressure of 580. mmHg and a temperature of 227 °C. What is the new pressure, in mmHg, of the gas when the temperature drops to 27 °C (V and n constant)?

   c. A spray can contains a gas with a pressure of 3.0 atm at a temperature of 17 °C. What is the pressure, in atmospheres, inside the container if the temperature rises to 110 °C (V and n constant)?

   d. A gas has a pressure of 1200. mmHg at 300. °C. What will the temperature, in degrees Celsius, be when the pressure falls to 1.10 atm (V and n constant)?

Answers   **a.** 2580 mmHg   **b.** 348 mmHg   **c.** 4.0 atm   **d.** 126 °C

6.6 The Combined Gas Law

• At constant amount of a gas, the gas laws can be combined into a relationship of pressure (P), volume (V), and temperature (T):

$$\frac{P_1 V_1}{T_1} = \frac{P_2 V_2}{T_2}$$ Subscripts 1 and 2 represent initial and final conditions.

(MC)

Tutorial: The Combined Gas Law

◆ **Learning Exercise 6.6**

Solve each of the following using the combined gas law (n is constant):

   a. A 5.00-L sample of nitrogen gas has a pressure of 1200. mmHg at 220. K. What is the pressure, in mmHg, of the sample when the volume increases to 20.0 L at 440. K?

b. A 25.0-mL bubble forms at the ocean depths where the pressure is 10.0 atm and the temperature is 5.0 °C. What is the volume, in milliliters, of that bubble at the ocean surface, where the pressure is 760.0 mmHg and the temperature is 25 °C?

c. A 35.0-mL sample of argon gas has a pressure of 1.0 atm and a temperature of 15 °C. What is the final volume, in milliliters, if the pressure goes to 2.0 atm and the temperature to 45 °C?

d. A weather balloon with a volume of 315-L is launched at the Earth's surface where the temperature is 12 °C and the pressure is 0.930 atm. What is the volume, in liters, of the balloon in the upper atmosphere where the pressure is 116 mmHg and the temperature is −35 °C?

e. A 10.0-L sample of gas is emitted from a volcano with a pressure of 1.20 atm and a temperature of 150. °C. What is the volume, in liters, of the gas when its pressure is 0.900 atm and the temperature is −40. °C?

Answers   **a.** 600. mmHg   **b.** 268 mL   **c.** 19 mL   **d.** 1.60×10^3 L   **e.** 7.34 L

6.7 Volume and Moles (Avogadro's Law)

- If the number of moles of gas increases, the volume increases; if the number of moles of gas decreases, the volume decreases.
- Avogadro's law states that equal volumes of gases at the same temperature and pressure contain the same number of moles. The volume (V) of a gas is directly related to the number of moles of the gas when the pressure and temperature of the gas do not change:

$$\frac{V_1}{n_1} = \frac{V_2}{n_2}$$ Subscripts 1 and 2 represent initial and final conditions.

- At STP conditions, standard pressure (1 atm) and temperature (0 °C), one mole of a gas occupies a volume of 22.4 L, the *molar volume*.

Guide to Using Molar Volume	
Step 1	State the given and needed quantities.
Step 2	Write a plan.
Step 3	Write conversion factors including 22.4 L/mole at STP.
Step 4	Set up the problem with factors to cancel units.

♦ Learning Exercise 6.7A

Fill in the blanks by writing I (increases) or D (decreases) for a gas in a closed container.

	Pressure	Volume	Moles	Temperature
a.	_____	Increases	Constant	Constant
b.	Increases	Constant	_____	Constant
c.	Constant	Decreases	_____	Constant
d.	_____	Constant	Constant	Increases
e.	Constant	_____	Constant	Decreases
f.	_____	Constant	Increases	Constant

Answers   a. D   b. I   c. D
              d. I   e. D   f. I

Study Note

At STP, the molar volume factor 22.4 L/mole of gas converts between moles of a gas and volume.

Example: What volume, in liters, would 2.00 moles of N_2 occupy at STP?

Step 1: Given 2.00 moles of N_2 **Need** liters of N_2
Step 2: Plan moles of N_2 → liters of N_2
Step 3: Conversion Factors

1 mole of gas = 22.4 L at STP

$$\frac{1 \text{ mole gas}}{22.4 \text{ L at STP}} \text{ and } \frac{22.4 \text{ L at STP}}{1 \text{ mole gas}}$$

Step 4: Set Up Problem

$$2.00 \text{ moles } N_2 \times \frac{22.4 \text{ L at STP}}{1 \text{ mole } N_2} = 44.8 \text{ moles of } N_2 \text{ at STP}$$

(MC)

Tutorial: Volume and Moles

♦ Learning Exercise 6.7B

Use Avogadro's law to solve the following gas problems:

a. A balloon containing 0.50 mole of helium has a volume of 4.00 L. What is the new volume, in liters, when an additional 1.0 mole of nitrogen is added to the balloon (*P* and *T* constant)?

b. A balloon containing 3.00 moles of helium has a volume of 15 L. What is the new volume, in liters, of the balloon when 2.00 moles of helium escapes from the balloon (*P* and *T* constant)?

c. What is the volume, in liters, occupied by 210. g of nitrogen (N_2) at STP (*P* and *T* constant)?

d. What is the volume, in liters, of 6.40 g of O_2 at STP?

Answers   **a.** 12 L   **b.** 5.0 L   **c.** 168 L   **d.** 4.48 L

6.8 Partial Pressures (Dalton's Law)

- In a mixture of two or more gases, the total pressure is the sum of the partial pressures (subscripts 1, 2, 3... represent the partial pressures of the individual gases).

$$P_{total} = P_1 + P_2 + P_3 + \dots$$

- The partial pressure of a gas in a mixture is the pressure it would exert if it were the only gas in the container.

Guide to Solving for Partial Pressure	
Step 1	Write the equation for the sum of partial pressures.
Step 2	Solve for the unknown quantity.
Step 3	Substitute known pressures and calculate the unknown partial pressure.

Tutorial: Mixtures of Gases

◆ **Learning Exercise 6.8A**

Use Dalton's law to solve the following problems about gas mixtures:

a. What is the pressure, in mmHg, of a sample of gases that contains oxygen at 0.500 atm, nitrogen (N_2) at 132 torr, and helium at 224 mmHg?

b. What is the pressure (atm) of a gas sample that contains helium at 285 mmHg and oxygen (O_2) at 1.20 atm?

c. A gas sample containing nitrogen (N_2) and oxygen (O_2) has a pressure of 1500. mmHg. If the partial pressure of the nitrogen is 0.900 atm, what is the partial pressure, in mmHg, of the oxygen gas in the mixture?

Answers   **a.** 736 mmHg   **b** 1.58 atm   **c.** 816 mmHg

♦ Learning Exercise 6.8B

Complete the following table for typical blood gas values for partial pressures:

Gas	Alveoli	Oxygenated blood	Deoxygenated blood	Tissues
O_2	_____	_____	_____	_____
CO_2	_____	_____	_____	_____

Answers

Gas	Alveoli	Oxygenated blood	Deoxygenated blood	Tissues
O_2	100 mmHg	100 mmHg	40 mmHg or less	30 mmHg or less
CO_2	40 mmHg	40 mmHg	46 mmHg	50 mmHg or greater

Checklist for Chapter 6

You are ready to take the Practice Test for Chapter 6. Be sure you have accomplished the following learning goals for this chapter. If you are not sure, review the section listed at the end of the goal. Then apply your new skills and understanding to the Practice Test.

After studying Chapter 6, I can successfully:

_____ Describe the kinetic molecular theory of gases (6.1).

_____ Change the units of pressure from one to another (6.2).

_____ Use the pressure–volume relationship (Boyle's law) to determine the new pressure or volume of a fixed amount of gas at constant temperature (6.3).

_____ Use the temperature–volume relationship (Charles's law) to determine the new temperature or volume of a fixed amount of gas at a constant pressure (6.4).

_____ Use the temperature–pressure relationship (Gay-Lussac's law) to determine the new temperature or pressure of a certain amount of gas at a constant volume (6.5).

_____ Use the combined gas law to find the new pressure, volume, or temperature of a certain amount of gas when changes in two of these properties are given (6.6).

_____ Describe the relationship between the amount of a gas and its volume (Avogadro's law) at constant *P* and *T*, and use this relationship in calculations (6.7).

_____ Use molar volume to convert the volume of gas at STP to the number of moles or grams of that gas at STP. Convert moles or grams of a gas at STP to volume at STP (6.7).

_____ Calculate the total pressure of a gas mixture from the partial pressures (6.8).

Practice Test for Chapter 6

For questions 1 through 5, answer using T (true) or F (false):

1. _____ A gas does not have its own volume or shape.

2. _____ The molecules of a gas are moving extremely fast.

3. _____ The collisions of gas molecules with the walls of their container create pressure.

4. _____ Gas molecules are close together and move in straight lines.

5. _____ Gas molecules have minimal attractions between them.

6. When a gas is heated in a closed metal container, the
   A. pressure increases   B. pressure decreases   C. volume increases
   D. volume decreases   E. number of molecules increases

7. The pressure of a gas will increase when
   A. the volume increases
   B. the temperature decreases
   C. more molecules of gas are added
   D. molecules of gas are removed
   E. none of these

8. If the temperature of a gas is increased,
   A. the pressure will decrease
   B. the volume will increase
   C. the volume will decrease
   D. the number of molecules will increase
   E. none of these

9. The relationship that the volume of a gas is inversely related to its pressure at constant temperature is known as
   A. Boyle's law   B. Charles's law   C. Gay-Lussac's law
   D. Dalton's law   E. Avogadro's law

10. What is the pressure (atm) of a gas with a pressure of 1200 mmHg?
  A. 0.63 atm **B.** 0.79 atm **C.** 1.2 atm
  D. 1.6 atm **E.** 2.0 atm

11. A 6.00-L sample of oxygen has a pressure of 660. mmHg. When the volume is reduced to 2.00 L at constant temperature and moles, it will have a new pressure of
  A. 1980 mmHg **B.** 1320 mmHg **C.** 330. mmHg
  D. 220. mmHg **E.** 110. mmHg

12. A sample of nitrogen gas at 110 K has a pressure of 1.0 atm. When the temperature is increased to 360 K at constant moles and volume, the new pressure will be
  A. 0.50 atm **B.** 1.0 atm **C.** 1.5 atm
  D. 3.3 atm **E.** 4.0 atm

13. If two gases have the same volume, temperature, and pressure, they also have the same
  A. density **B.** number of molecules **C.** molar mass
  D. speed **E.** size molecules

14. A gas sample with a volume of 4.0 L has a pressure of 750 mmHg and a temperature of 77 °C. What is its new volume at 277 °C and 250 mmHg (*n* constant)?
  A. 7.6 L **B.** 19 L **C.** 2.1 L
  D. 0.00056 L **E.** 3.3 L

15. If the temperature and amount of a gas do not change, but its volume doubles, its pressure will
  A. double
  B. triple
  C. decrease to one-half the original pressure
  D. decrease to one-fourth the original pressure
  E. not change

16. A sample of oxygen with a pressure of 600. mmHg at 127 °C has a volume of 4.00 L. What will the new pressure, in mmHg, be when the volume expands to 12.0 L while temperature changes to 227 °C?
  A. 160. mmHg **B.** 250. mmHg **C.** 357 mmHg
  D. 800. mmHg **E.** 1560. mmHg

17. A sample of 2.00 moles of gas initially at STP is converted to a volume of 5.0 L and a temperature of 27 °C. What is its new pressure in atmospheres?
  A. 0.12 atm **B.** 5.5 atm **C.** 7.5 atm
  D. 9.8 atm **E.** 10. atm

18. The conditions for standard temperature and pressure (STP) are
  A. 0 K, 1 atm **B.** 0 °C, 10 atm **C.** 25 °C, 1 atm
  D. 273 K, 1 atm **E.** 273 K, 0.5 atm

19. The volume occupied by 1.50 moles of CH_4 at STP is
  A. 44.8 L **B.** 33.6 L **C.** 22.4 L
  D. 11.2 L **E.** 5.60 L

20. How many grams of oxygen gas (O_2) are present in 44.1 L of oxygen gas at STP?
  A. 10.0 g **B.** 16.0 g **C.** 32.0 g
  D. 410.0 g **E.** 63.0 g

21. What is the volume, in liters, of 0.50 mole of nitrogen gas (N_2) at 25 °C and 2.0 atm?
  A. 0.51 L   **B.** 1.0 L   **C.** 4.2 L
  D. 6.1 L   **E.** 24 L

22. A gas mixture contains helium with a partial pressure of 0.100 atm, oxygen with a partial pressure of 445 mmHg, and nitrogen gas with a partial pressure of 235 mmHg. What is the total pressure, in atm, for the gas mixture?
  A. 0.995 atm   **B.** 1.39 atm   **C.** 1.69 atm
  D. 2.00 atm   **E.** 10.0 atm

23. A mixture of oxygen gas and nitrogen gas has a total pressure of 1040 mmHg. If the oxygen gas has a partial pressure of 510 mmHg, what is the partial pressure of the nitrogen gas?
  A. 240 mmHg   **B.** 530 mmHg   **C.** 775 mmHg
  D. 1040 mmHg   **E.** 1350 mmHg

24. 3.00 moles of He in a steel container has a pressure of 12.0 atm. What is the new pressure after 4.00 moles of He is added (*T* and *V* remain constant)?
  A. 5.14 atm   **B.** 16.0 atm   **C.** 28.0 atm
  D. 32.0 atm   **E.** 45.0 atm

25. The exchange of gases between the alveoli, blood, and tissues of the body is a result of
  A. pressure gradients   **B.** different molecular masses
  C. shapes of molecules   **D.** altitude
  E. all of these

26. Oxygen moves into the tissues from the blood because its partial pressure
  A. in arterial blood is higher than in the tissues
  B. in venous blood is higher than in the tissues
  C. in arterial blood is lower than in the tissues
  D. in venous blood is lower than in the tissues
  E. is equal in the blood and in the tissues

Answers to the Practice Test

1. T	**2.** T	**3.** T	**4.** F	**5.** T
6. A	**7.** C	**8.** B	**9.** A	**10.** D
11. A	**12.** D	**13.** B	**14.** B	**15.** C
16. B	**17.** D	**18.** D	**19.** B	**20.** E
21. D	**22.** A	**23.** B	**24.** C	**25.** A
26. A				

Solutions

Study Goals

- Identify the solute and solvent in a solution.
- Describe the formation of a solution.
- Identify solutes as electrolytes and nonelectrolytes and distinguish between strong electrolytes, weak electrolytes, and nonelectrolytes in solution.
- Define solubility, distinguish between a saturated and unsaturated solution, and determine whether a salt will dissolve in water.
- Calculate the mass percent (m/m), volume percent (v/v), and mass/volume percent (m/v) concentrations of a solution; use percent concentration as a conversion factor to determine mass or volume of solute or mass or volume of solution.
- Calculate the molarity of a solution; use molarity as a conversion factor to determine moles of solute or volume of solution.
- Describe how to prepare a diluted solution.
- Identify a mixture as a solution, a colloid, or a suspension.
- Describe how number of particles affects the freezing point, boiling point, and osmotic pressure.
- Describe osmosis and isotonic, hypotonic, and hypertonic solutions.
- Describe the behavior of a red blood cell in hypotonic, isotonic, and hypertonic solutions.

Think About It

1. Why is tea called a solution, but the water used to make tea is a pure substance?

2. Sugar dissolves in water, but oil does not. Why?

3. Why is salt used in the preparation of ice cream?

4. Why are pickles made in a salt solution or "brine" with a high salt concentration?

5. How do your kidneys remove toxic substances from the blood but retain the usable substances?

Key Terms

Match each of the following key terms with the correct definitions.

    a. solution   **b.** mass percent (m/m) concentration   **c.** molarity

    d. osmosis   **e.** strong electrolyte   **f.** solubility

1. _____ a substance that completely dissociates into ions when it dissolves in water

2. _____ the grams of solute dissolved in 100 g of solution

3. _____ the number of moles of solute in 1 L of solution

4. _____ the flow of a solvent through a semipermeable membrane into a solution with a higher solute concentration

5. _____ a mixture of a solute and a solvent

6. _____ the maximum amount of a solution that can dissolve under given conditions

Answers   **1.** e   **2.** b   **3.** c   **4.** d   **5.** a   **6.** f

7.1 Solutions

- A solution is a homogeneous mixture that forms when a solute dissolves in a solvent. The solvent is present in a greater amount.
- Solvents and solutes can be solids, liquids, or gases. The solution is the same physical state as the solvent.
- A polar solute is soluble in a polar solvent; a nonpolar solute is soluble in a nonpolar solvent.
- Water molecules form hydrogen bonds because the partial positive charge of the hydrogen in one water molecule is attracted to the partial negative charge of oxygen in another water molecule.
- An ionic solute dissolves in water, a polar solvent, because the polar water molecules attract and pull the positive and negative ions into solution. In solution, water molecules surround the ions in a process called *hydration*.

(MC)

Self Study Activity: Hydrogen Bonding

♦ **Learning Exercise 7.1A**

Water is polar, and hexane is nonpolar. In which solvent is each of the following soluble?

    a. bromine, Br_2, nonpolar _____   **b.** HCl, polar   _____

    c. cholesterol, nonpolar _____   **d.** vitamin D, nonpolar _____

    e. vitamin C, polar _____

Answers   **a.** hexane   **b.** water   **c.** hexane   **d.** hexane   **e.** water

♦ **Learning Exercise 7.1B**

Indicate the solute and solvent in each of the following: **Solute**   **Solvent**

    a. 10 g of KCl dissolved in 100 g of water   _____   _____

    b. soda water: $CO_2(g)$ dissolved in water   _____   _____

    c. an alloy composed of 80% Zn and 20% Cu   _____   _____

    d. a mixture of O_2 (200 mmHg) and He (500 mmHg)   _____ _____

    e. a solution of 40 mL of CCl_4 and 2 mL of Br_2 _____ _____

Answers   **a.** KCl (solute); water (solvent)   **b.** CO_2 (solute); water (solvent)
                    c. Cu (solute); Zn (solvent)   **d.** O_2 (solute); He (solvent)
                    e. Br_2 (solute); CCl_4 (solvent)

7.2 Electrolytes and Nonelectrolytes

- Electrolytes conduct an electrical current because they produce ions in aqueous solutions.
- Strong electrolytes are completely ionized, whereas weak electrolytes are slightly ionized.
- Nonelectrolytes do not form ions in solution but dissolve as molecules.
- An equivalent is the amount of an electrolyte that carries 1 mole of electrical charge.
- The number of equivalents per mole of a positive or negative ion is equal to the charge on the ion.

♦ **Learning Exercise 7.2A**

Write an equation for the formation of an aqueous solution of each of the following strong electrolytes:

    a. LiCl

    b. $Mg(NO_3)_2$

    c. Na_3PO_4

    d. K_2SO_4

    e. $MgCl_2$

Answers   **a.** $LiCl(s) \xrightarrow{H_2O} Li^+(aq) + Cl^-(aq)$

                  b. $Mg(NO_3)_2(s) \xrightarrow{H_2O} Mg^{2+}(aq) + 2NO_3^-(aq)$

                  c. $Na_3PO_4(s) \xrightarrow{H_2O} 3Na^+(aq) + PO_4^{3-}(aq)$

                  d. $K_2SO_4(s) \xrightarrow{H_2O} 2K^+(aq) + SO_4^{2-}(aq)$

                  e. $MgCl_2(s) \xrightarrow{H_2O} Mg^{2+}(aq) + 2Cl^-(aq)$

♦ **Learning Exercise 7.2B**

Indicate whether an aqueous solution of each of the following contains ions only, molecules only, or mostly molecules with some ions. Write an equation for the formation of the solution.

a. glucose, $C_6H_{12}O_6$, a nonelectrolyte

b. NaOH, a strong electrolyte

c. K_2SO_4, a strong electrolyte

d. HF, a weak electrolyte

Answers
  a. $C_6H_{12}O_6(s) \xrightarrow{H_2O} C_6H_{12}O_6(aq)$ molecules only
  b. $NaOH(s) \xrightarrow{H_2O} Na^+(aq) + OH^-(aq)$ ions only
  c. $K_2SO_4(s) \xrightarrow{H_2O} 2K^+(aq) + SO_4{}^{2-}(aq)$ ions only
  d. $HF(aq) \underset{\longleftarrow}{\xrightarrow{H_2O}} H^+(aq) + F^-(aq)$ mostly molecules and a few ions

♦ **Learning Exercise 7.2C**

Calculate the following:

a. Number of equivalents in 1 mole of Mg^{2+}

b. Number of equivalents of Cl^- in 2.5 moles of Cl^-

c. Number of equivalents of Ca^{2+} in 2.0 moles of Ca^{2+}

Answers   **a.** 2 Eq   **b.** 2.5 Eq   **c.** 4.0 Eq

7.3 Solubility

- The amount of solute that dissolves depends on the nature of the solute and solvent.
- Solubility describes the maximum amount of a solute that dissolves in exactly 100 g of solvent at a given temperature.
- A saturated solution contains the maximum amount of dissolved solute at a certain temperature whereas an unsaturated solution contains less than this amount.
- An increase in temperature increases the solubility of most solids, but decreases the solubility of gases in water.

- Henry's law states that the solubility of gases in liquids is directly related to the partial pressure of that gas over the liquid.
- The solubility rules describe the ionic combinations that are soluble and insoluble in water. If a salt contains an NH_4^+, Li^+, Na^+, K^+, NO_3^-, or $C_2H_3O_2^-$ ion, it is soluble in water. Most halides and sulfates are soluble.

(MC)

Self Study Activity: Solubility
Tutorial: Solubility
Tutorial: Solubility of Gases and Solids in Water
Tutorial: Electrolytes and Ionization
Case Study: Kidney Stones and Saturated Solutions

♦ **Learning Exercise 7.3A**

Identify each of the following as a (S) saturated solution or an (U) unsaturated solution:

1. A sugar cube dissolves when added to a cup of coffee. _____

2. A KCl crystal added to a KCl solution does not change in size. _____

3. A layer of sugar forms in the bottom of a glass of tea. _____

4. The rate of crystal formation is equal to the rate of solution. _____

5. Upon heating, all the sugar in a solution dissolves. _____

Answers　　　1. U　　　2. S　　　3. S　　　4. S　　　5. U

♦ **Learning Exercise 7.3B**

Use the solubility table for $NaNO_3$ to answer each of the following problems:

Temperature (°C)	Solubility $\left(g \text{ of } NaNO_3/100. g \text{ of } H_2O\right)$
40	100.
60	120.
80	150.
100	200.

　　a. How many grams of $NaNO_3$ will dissolve in 100. g water at 40 °C?

　　b. How many grams of $NaNO_3$ will dissolve in 300. g of water at 60 °C?

c. A mixture is prepared using 200. g of water and 350. g of $NaNO_3$ at 80 °C. Will any solute remain undissolved? If so how much?

d. Will 150. g of $NaNO_3$ all dissolve when added to 100. g of water at 100 °C?

Answers   **a.** 100. g
           b. 360. g
           c. 300. g of $NaNO_3$ will dissolve, leaving 50. g of $NaNO_3$ undissolved.
           d. Yes. All 150. g of $NaNO_3$ will dissolve.

♦ **Learning Exercise 7.3C**

Predict whether the following salts are soluble (S) or insoluble (I) in water:

1. _____ NaCl   **2.** _____ $AgNO_3$ **3.** _____ $PbCl_2$

4. _____ Ag_2S **5.** _____ $BaSO_4$ **6.** _____ Na_2CO_3

7. _____ K_2S **8.** _____ $MgCl_2$ **9.** _____ BaS

Answers   **1.** S   **2.** S   **3.** I   **4.** I   **5.** I
           6. S   **7.** S   **8.** S   **9.** I

7.4 Concentration of a Solution

- The concentration of a solution is the relationship between the amount of solute (g or mL) and the amount (g or mL) of solution.
- A mass percent (m/m) concentration expresses the ratio of the mass of solute to the mass of solution multiplied by 100.

$$\text{Percent (m/m)} = \frac{\text{mass of solute (g)}}{\text{mass of solution (g)}} \times 100\%$$

- A volume percent (v/v) concentration expresses the ratio of the volume of the solute to the volume of the solution multiplied by 100%.

$$\text{Percent (v/v)} = \frac{\text{volume of solute (mL)}}{\text{volume of solution (mL)}} \times 100\%$$

- A mass/volume percent (m/v) concentration expresses the ratio of the mass of the solute to the volume (mL) of the solution multiplied by 100%.

$$\text{Percent (m/v)} = \frac{\text{mass of solute (g)}}{\text{volume of solution (mL)}} \times 100\%$$

- A molarity (M) concentration expresses the number of moles of solute dissolved in 1 L (1000 mL) of a solution.

$$\text{Molarity (M)} = \frac{\text{moles solute}}{\text{liters solution}}$$

Guide to Calculating Solution Concentration	
Step 1	Determine the quantities of solute and solution.
Step 2	Write the concentration expression.
Step 3	Substitute solute and solution quantities into the expression.

Study Note

Example: What is the percent (mass/mass) when 2.4 g of $NaHCO_3$ dissolves in 120 g of solution?

Step 1: Given 2.4 g of NaHCO3; 120 g of solution   **Need** mass percent (m/m) concentration

Step 2: Write the concentration expression.

$$\text{mass percent (m/m)} = \frac{\text{mass of solute}}{\text{mass of solution}} \times 100\% \text{ mass percent}$$

Step 3: Substitute solute and solution quantities.

$$\text{mass percent (m/m)} = \frac{2.4 \text{ g } NaHCO_3}{120 \text{ g solution}} \times 100\% = 2.0\% \text{ (m/m)}$$

(MC)

Tutorial: Calculating Percent Concentration
Tutorial: Percent Concentration as a Conversion Factor

♦ **Learning Exercise 7.4A**

Determine the percent concentration for each of the following solutions:

a. mass percent (m/m) for 18.0 g of NaCl in 90.0 g of NaCl solution

b. mass percent (m/v) for 5.0 g of KCl in 2.0 L of KCl solution

c. mass/volume percent (m/v) for 4.0 g of KOH in 50.0 mL of KOH solution

d. volume percent (v/v) for 21 mL of alcohol in 350 mL of mouthwash solution

Answers   **a.** 20.0%   **b.** 0.25%   **c.** 8.0%   **d.** 6.0%

♦ **Learning Exercise 7.4B**

Calculate the molarity of each of the following solutions:

a. 1.50 moles of HCl in 0.500 L of HCl solution

b. 10.0 moles of glucose ($C_6H_{12}O_6$) in 2.50 L of glucose solution

c. 80.0 g of NaOH in 4.0 L of NaOH solution (*Hint*: Find moles of NaOH.)

d. 53 g of Na_2CO_3 in 250 mL of Na_2CO_3 solution

Answers   **a.** 3.00 M HCl solution   **b.** 4.00 M glucose solution
              c. 0.50 M NaOH   **d.** 2.0 M Na_2CO_3

Guide to Using Concentration to Calculate Mass or Volume	
Step 1	State the given and needed quantities.
Step 2	Write a plan to calculate mass or volume.
Step 3	Write equalities and conversion factors.
Step 4	Set up the problem to calculate mass or volume.

Study Note

Example: How many grams of KI are needed to prepare 225 g of a 4.0% (m/m) KI solution?

Step 1: State the given and needed quantities.

Given 225 g of a 4.0 % (m/m) KI solution

Need grams of KI

Step 2: Write a plan to calculate mass or volume.

grams of KI solution → grams of KI (solute)

Step 3: Write equalities and conversion factors.

100 g of KI solution = 4.0 g of KI (solute)

$$\frac{4.0 \text{ g KI}}{100 \text{ g KI solution}} \quad \text{and} \quad \frac{100 \text{ g KI solution}}{4.0 \text{ g KI}}$$

Step 4: Set up the problem to calculate mass or volume.

$$225 \text{ g KI solution} \times \frac{4.0 \text{ g KI}}{100 \text{ g KI solution}} = 9.0 \text{ g of KI}$$

♦ **Learning Exercise 7.4C**

Calculate the number of grams of solute needed to prepare each of the following solutions:

a. grams of glucose to prepare 400. g of a 5.0% (m/m) solution

b. grams of lidocaine hydrochloride to prepare 50.0 g of a 2.0% (m/m) solution

c. grams of KCl to prepare 1.2 L g of a 4.0% (m/v) KCl solution

d. grams of NaCl to prepare 1.50 L of a 2.00% (m/v) NaCl solution

Answers   **a.** 20. g   **b.** 1.0 g   **c.** 48 g   **d.** 30.0 g

♦ **Learning Exercise 7.4D**

Use percent concentration to calculate the specified quantity of each solution:

   a. How many milliliters of a 1.00% (m/v) NaCl solution contain 2.00 g of NaCl?

   b. How many milliliters of a 5.0% (m/v) glucose solution contain 24 g of glucose?

   c. How many milliliters of a 2.5% (v/v) methanol solution contain 7.0 mL of methanol?

   d. How many grams of a 15% (m/m) NaOH solution contain 7.5 g of NaOH?

Answers   **a.** 200. mL   **b.** 480 mL   **c.** 280 mL   **d.** 50. g

---

Study Note

Molarity is used as a conversion factor to convert between the amount of solute and the volume of solution. How many grams of NaOH are in 0.250 L of a 5.00 M NaOH solution?

   Step 1: State the given and needed quantities.

   Given 0.250 L of a 5.00 M NaOH solution

   Need grams of NaOH

   Step 2: Write a plan to calculate mass or volume.

   Liters of solution → moles of solution → grams of solute

   Step 3: Write equalities and conversion factors. The concentration 5.00 M can be expressed as conversion factors.

1 L of NaOH solution = 5.00 moles of NaOH   1 mole of NaOH = 40.0 g of NaOH

$$\frac{5.00 \text{ moles NaOH}}{1 \text{ L NaOH solution}} \text{ and } \frac{1 \text{ L NaOH solution}}{5.00 \text{ moles NaOH}} \qquad \frac{1 \text{ mole NaOH}}{40.0 \text{ g NaOH}} \text{ and } \frac{40.0 \text{ g of NaOH}}{1 \text{ mole NaOH}}$$

   Step 4: Set up problem to calculate mass or volume.

$$0.250 \text{ L solution} \times \frac{5.00 \text{ moles NaOH}}{1 \text{ L solution}} \times \frac{40.0 \text{ g NaOH}}{1 \text{ mole NaOH}} = 50.0 \text{ g of NaOH}$$

♦ **Learning Exercise 7.4E**

Calculate the quantity of solute in each of the following solutions:

  a. How many moles of HCl are in 1.50 L of a 6.00 M HCl solution?

  b. How many mole of KOH are in 0.750 L of a 10.0 M KOH solution?

  c. How many grams of NaOH are needed to prepare 0.500 L of a 4.40 M NaOH solution? (*Hint*: Find moles of NaOH.)

  d. How many grams of NaCl are in 285 mL of a 1.75 M NaCl solution?

Answers   a. 9.00 moles of HCl   b. 7.50 moles of KOH   c. 88.0 g of NaOH
         d. 29.2 g of NaCl

♦ **Learning Exercise 7.4F**

Calculate the milliliters needed of each solution to obtain the given quantity of solute:

  a. 1.50 moles of $Mg(OH)_2$ from a 2.00 M $Mg(OH)_2$ solution

  b. 0.150 mole of glucose from a 2.20 M glucose solution

c. 18.5 g of KI from a 3.00 M KI solution

d. 18.0 g of NaOH from a 6.00 M NaOH solution

Answers   **a.** 750. mL   **b.** 68.2 mL   **c.** 37.1 mL   **d.** 75.0 mL

7.5 Dilution of Solutions

- *Dilution* is the process of mixing a solution with solvent to obtain a lower concentration.
- For dilutions, solve for the unknown value in the expression $C_1V_1 = C_2V_2$ or $M_1V_1 = M_2V_2$ (subscript 1 represents the concentrated solution and subscript 2 represents the diluted solution).

Tutorial: Solution Dilution

Guide to Calculating Dilution Quantities	
Step 1	Prepare a table of the concentrations and volumes of the solutions.
Step 2	Rearrange the dilution expression for the unknown quantity.
Step 3	Substitute the unknown quantities into the dilution expression and solve.

Study Note

What is the final concentration after 150. mL of a 2.00 M NaCl solution is diluted to a volume of 400. mL?
Solution

Step 1: Prepare a table of the concentrations and volumes of the solutions.

Concentrated Solution	Diluted Solution	Know	Predict
$M_1 = 2.00$ M	$M_2 = ?$		M decreases
$V_1 = 150.$ mL	$V_2 = 400.$ mL	V increases	

Step 2: Rearrange the dilution expression for the unknown quantity.

$$M_1V_1 = M_2V_2 \qquad M_2 = M_1 \times \frac{V_1}{V_2}$$

Step 3: Substitute the unknown quantities into the dilution expression and solve.

$$M_2 = 2.00 \text{ M} \times \frac{150.\ \text{mL}}{400.\ \text{mL}} = 0.750 \text{ M}$$

♦ Learning Exercise 7.5

Solve each of the following dilution problems (assume the volumes add):

a. What is the final concentration after 100. mL of a 5.0 M KCl solution is diluted with water to give a final volume of 200. mL?

b. What is the final concentration of the diluted solution if 5.0 mL of a 15.% (m/v) KCl solution is diluted to 25 mL?

c. What is the final concentration after 250 mL of an 8.0 M NaOH solution is diluted with 750 mL of water?

d. 160. mL of water is added to 40. mL of a 1.0 M NaCl solution. What is the final concentration?

e. What volume of 6.0% (m/v) HCl is needed to prepare 300. mL of a 1.0% (m/v) HCl solution? How much water must be added?

Answers　　　**a.** 2.5 M　　　　**b.** 3.0%　　　　**c.** 2.7 M　　　　**d.** 0.20 M
　　　　　　　　e. $V_1 = 50.$ mL; add 250. mL of water

7.6 Properties of Solutions

- Colloids are homogeneous mixtures that contain solute particles that do not settle out. The particles will pass through a filter but not through a semipermeable membrane.
- Suspensions are composed of large particles that settle out of solution.
- The freezing point of a solution is lower than that of the solvent and boiling point is higher. The change in boiling or freezing point depends only on the number of particles of solute in the solution.
- In the process of osmosis, water (solvent) moves through a semipermeable membrane from the solution that has a lower solute concentration to a solution where the solute concentration is higher.

- Osmotic pressure is the pressure that prevents the flow of water into a more concentrated solution.
- Isotonic solutions have osmotic pressures equal to that of body fluids. A hypotonic solution has a lower osmotic pressure than body fluids; a hypertonic solution has a higher osmotic pressure.
- A red blood cell maintains its volume in an isotonic solution, but it swells (hemolysis) in a hypotonic solution and shrinks (crenation) in a hypertonic solution.
- In dialysis, water and small solute particles can pass through a dialyzing membrane, while larger particles such as blood cells are retained.

(MC)

Tutorial: Osmosis
Tutorial: Dialysis

◆ Learning Exercise 7.6A

Identify each of the following as a solution, colloid, or suspension:

1. contains single atoms, ions, or small molecules

2. contains particles that settle out

3. contains particles that are retained by filters

4. contains particles that move through a cellular membrane

5. contains particles retained by cellular membranes but not by filters

Answers	1. solution	2. suspension	3. suspension
	4. solution	5. colloid	

◆ Learning Exercise 7.6B

Fill in the blanks:
In osmosis, the direction of solvent flow is from the (1) [higher/lower] solute concentration to the (2) [higher/lower] solute concentration. A semipermeable membrane separates 5% (m/v) and 10% (m/v) sucrose solutions. The (3) _____% (m/v) solution has the greater osmotic pressure. Water will move from the (4) _____% (m/v) solution into the (5) _____% (m/v) solution. The compartment that contains the (6) _____% (m/v) solution increases in volume.

Answers	(1) lower	(2) higher	(3) 10
	(4) 5	(5) 10	(6) 10

◆ Learning Exercise 7.6C

A semipermeable membrane separates a 2% starch solution from a 10% starch solution. Complete each of the following with 2% or 10%:

a. Water will flow from the _____ starch solution to the _____ starch solution.

b. The volume of the _____ starch solution will increase and the volume of the _____ starch solution will decrease.

Answers	a. 2%, 10%	b. 10%, 2%

♦ **Learning Exercise 7.6D**

An aqueous solution with dissolved particles will have a higher boiling point and a lower freezing point than pure water. A solution of a strong electrolyte will raise the boiling point or lower the freezing point more than a solution of a nonelectrolyte having the same concentration. For example, a 1.0 L solution of 1 mole of NaCl, a strong electrolyte, produces 2 moles of particles (Na^+ and Cl^-) whereas a 1.0 L solution of 1 mole of glucose, a nonelectrolyte, produces 1 mole of glucose particles. Thus the freezing point of the NaCl solution would be lowered twice as much as that of the glucose solution. The boiling point for the NaCl solution would be raised twice as much as that of the glucose solution. For each of the following solutes in 1.0 L of water determine:

1. the number of moles of particles for each solute

2. the solute that would produce the largest change in freezing point and boiling point

3. the solute that would produce the smallest change in freezing point and boiling point

Solute	Number of Moles of Particles	Produces the Largest Change in Freezing Point/Boiling Point	Produces the Smallest Change in Freezing Point/Boiling Point
a. 1.0 mole of fructose (nonelectrolyte)			
b. 1.0 mole of KCl (strong electrolyte)			
c. 1.0 mole of $Ca(NO_3)_2$ (strong electrolyte)			

Answers

Solute	Number of Moles of Particles	Produces the Largest Change in Freezing Point/Boiling Point	Produces the Smallest Change in Freezing Point/Boiling Point
a. 1.0 mole of fructose (nonelectrolyte)	1.0 mole		1.0 mole of fructose
b. 1.0 mole of KCl (strong electrolyte)	2.0 mole		
c. 1.0. mole of $Ca(NO_3)_2$ (strong electrolyte)	3.0 mole	1.0 mole of $Ca(NO_3)_2$	

♦ **Learning Exercise 7.6E**

Fill in the blanks:
A (1) _____% (m/v) NaCl solution and a (2) _____% (m/v) glucose solution are isotonic to the body fluids. A red blood cell placed in these solutions does not change in volume because these solutions are (3) _____ tonic. When a red blood cell is placed in water, it undergoes (4) _____ because water is (5) _____ tonic. A 20% (m/v) glucose solution will cause a red blood cell to undergo (6) _____ because the 20% (m/v) glucose solution is (7) _____ tonic.

Answers (1) 0.9 (2) 5 (3) iso (4) hemolysis
  (5) hypo (6) crenation (7) hyper

♦ Learning Exercise 7.6F

Compared to body fluids, indicate whether the following solutions are

  a. hypotonic **b.** hypertonic **c.** isotonic

  1. 5% (m/v) glucose **2.** 3% (m/v) NaCl **3.** 2% (m/v) glucose

  4. water **5.** 0.9% (m/v) NaCl **6.** 10% (m/v) glucose

Answers **1.** c **2.** b **3.** a **4.** a **5.** c **6.** b

♦ Learning Exercise 7.6G

Indicate whether the following solutions will cause a red blood cell to undergo

  a. crenation **b.** hemolysis **c.** no change (stays the same)

  1. 10% (m/v) NaCl **2.** 1% (m/v) glucose **3.** 5% (m/v) glucose

  4. 0.5% (m/v) NaCl **5.** 10% (m/v) glucose **6.** water

Answers **1.** a **2.** b **3.** c **4.** b **5.** a **6.** b

♦ Learning Exercise 7.6H

A dialysis bag contains starch, glucose, NaCl, protein, and urea.

  a. When the dialysis bag is placed in water, what components would you expect to dialyze through the bag? Why?

  b. Which components will stay inside the dialysis bag? Why?

Answers **a.** Glucose, NaCl, and urea are solution particles. Solution particles will pass through semipermeable membranes.
  b. Starch and protein are colloids; colloids are retained by semipermeable membranes.

Checklist for Chapter 7

You are ready to take the Practice Test for Chapter 7. Be sure you have accomplished the following learning goals for this chapter. If you are not sure, review the section listed at the end of the goal. Then apply your new skills and understanding to the Practice Test.

After studying Chapter 7, I can successfully:

_____ Identify the solute and solvent in a solution and describe the process of dissolving an ionic solute in water (7.1).

_____ Identify the components in solutions of electrolytes and nonelectrolytes (7.2).

_____ Identify a saturated and an unsaturated solution (7.3).

_____ Identify a salt as soluble or insoluble (7.3).

_____ Determine the solubility of an ionic compound (salt) in water (7.3).

_____ Calculate the percent concentration, m/m, v/v, or m/v of a solute in a solution, and use percent concentration as a conversion factor to calculate the amount of solute or solution (7.4).

_____ Calculate the molarity of a solution (7.4).

_____ Use molarity as a conversion factor to calculate between the moles (or grams) of a solute and the volume of the solution (7.4).

_____ Calculate the new concentration or new volume when a solution is diluted (7.5).

_____ Identify a solution as isotonic, hypotonic, or hypertonic (7.6).

_____ Identify a mixture as a solution, a colloid, or a suspension (7.6).

_____ Explain the processes of freezing point lowering, boiling point elevation, osmosis, and dialysis (7.6).

_____ Describe how the number of particles in a solution affects the freezing point, boiling point, or osmotic pressure of a solution (7.6).

Practice Test for Chapter 7

For questions 1 through 4, indicate if each of the following is more soluble in (W) water, a polar solvent, or (B) benzene, a nonpolar solvent:

1. $I_2(g)$, nonpolar _____   **2.** NaBr(s), polar _____

3. KI(s), polar _____   **4.** C_6H_{12}, nonpolar _____

5. When dissolved in water, $Ca(NO_3)_2(s)$ dissociates into

   A. $Ca^{2+}(aq)+(NO_3)_2^{2-}(aq)$ **B.** $Ca^+(aq)+NO_3^-(aq)$ **C.** $Ca^{2+}(aq)+2NO_3^-(aq)$

   D. $Ca^{2+}(aq)+2N^{5+}(aq)+2O_3^{6-}(aq)$ **E.** $CaNO_3^+(aq)+NO_3^-(aq)$

6. Ethanol, CH_3—CH_2—OH, is a nonelectrolyte. When placed in water it

   A. dissociates completely.   **B.** dissociates partially.   **C.** does not dissociate.
   D. makes the solution acidic.   **E.** makes the solution basic.

7. The solubility of NH_4Cl is 46 g in 100. g of water at 40 °C. How much NH_4Cl can dissolve in 500. g of water at 40 °C?

   A. 9.2 g   **B.** 46 g   **C.** 100 g   **D.** 184 g   **E.** 230 g

For questions 8 through 11, predict if each of the following are soluble (S) or insoluble (I) in water:

8. NaCl _____   **9.** AgCl _____   **10.** $BaSO_4$ _____   **11.** FeO _____

12. A solution containing 1.20 g of sucrose in 50.0 g of solution has a mass percent (m/m) concentration of
   A. 0.600%   **B.** 1.20%   **C.** 2.40%   **D.** 30.0%   **E.** 41.6%

13. The amount of lactose in 250 g of a 3.0% (m/m) lactose solution for infant formula is
   A. 0.15 g   **B.** 1.2 g   **C.** 6.0 g   **D.** 7.5 g   **E.** 30 g

14. The mass of solution of 5.0% (m/m) glucose solution that contains 0.40 g of glucose is
  A. 1.0 g   **B.** 2.0 g   **C.** 4.0 g   **D.** 5.0 g   **E.** 8.0 g

15. The mass of NaCl needed to prepare 50.0 g of a 4.00% (m/m) NaCl solution is
  A. 20.0 g   **B.** 15.0 g   **C.** 10.0 g   **D.** 4.00 g   **E.** 2.00 g

16. A solution containing 6.0 g of NaCl in 1500 g of solution has a mass percent (m/m) concentration of
  A. 0.40% (m/m)   **B.** 0.25% (m/m)   **C.** 4.0% (m/m)   **D.** 0.90% (m/m)   **E.** 2.5% (m/m)

17. The moles of KOH needed to prepare 2400 mL of a 2.0 M KOH solution is
  A. 1.2 moles   **B.** 2.4 moles   **C.** 4.8 moles   **D.** 12 moles   **E.** 48 moles

18. The grams of NaOH needed to prepare 7.5 mL of a 5.0 M NaOH is
  A. 1.5 g   **B.** 3.8 g   **C.** 6.7 g   **D.** 15 g   **E.** 38 g

For questions 19 through 21, consider a 20.0-g sample of a solution that contains 2.0 g of NaOH.

19. The mass percent (m/m) concentration of the solution is
  A. 1.0%   **B.** 4.0%   **C.** 5.0%   **D.** 10.%   **E.** 20.%

20. The moles of NaOH in the sample is
  A. 0.050 mole   **B.** 0.40 mole   **C.** 1.0 mole   **D.** 2.5 moles   **E.** 4.0 moles

21. If the solution has a volume of 0.025 L, what is the molarity of the sample?
  A. 0.10 M   **B.** 0.50 M   **C.** 1.0 M   **D.** 1.5 M   **E.** 2.0 M

22. Which of the following is soluble in water?
  A. $AgCl$ **B.** $BaCO_3$ **C.** K_2SO_4 **D.** PbS **E.** MgO

23. When the pressure of a gas above a solution is increased, the solubility of the gas in the solution will:
  A. increase   **B.** decrease   **C.** not change

24. Which of the following salts is insoluble in water?
  A. $CuCl_2$ **B.** $Pb(NO_3)_2$ **C.** K_2CO_3 **D.** $(NH_4)SO_4$ **E.** $CaCO_3$

25. A 20.-mL sample of 5.0 M HCl is diluted with water to give 100. mL of solution. The final concentration of the HCl solution is
  A. 10 M   **B.** 5.0 M   **C.** 2.0 M   **D.** 1.0 M   **E.** 0.50 M

26. Water is added to 200. mL of a 4.00 M KNO_3 solution to give 400. mL of solution. The final concentration of the diluted solution is
  A. 1.00 M   **B.** 2.00 M   **C.** 4.00 M   **D.** 0.500 M   **E.** 0.100 M

27. 5.0 mL of a 2.0 M KOH solution is diluted with water to give 50.0 mL of solution. The final concentration of the KOH solution is
  A. 1.5 M   **B.** 1.0 M   **C.** 20. M   **D.** 0.20 M   **E.** 0.50 M

For questions 28 through 32, indicate whether each statement describes a
  A. solution   **B.** colloid   **C.** suspension

28. contains single atoms, ions, or small molecules of solute

29. settles out upon standing

30. can be separated by filtering

31. can be separated by semipermeable membranes

32. passes through semipermeable membranes

33. Two solutions that have identical osmotic pressures are
    A. hypotonic   **B.** hypertonic   **C.** isotonic
    D. isotopic   **E.** hyperactive

34. In osmosis, the net flow of water is
    A. between solutions of equal concentrations
    B. from higher solute concentration to lower solute concentration
    C. from lower solute concentration to higher solute concentration
    D. from a colloid to a solution of equal concentration
    E. from lower solvent concentration to higher solvent concentration

35. A red blood cell undergoes hemolysis when placed in a solution that is
    A. isotonic   **B.** hypotonic   **C.** hypertonic
    D. colloidal   **E.** semitonic

36. A solution that has the same osmotic pressure as body fluids is
    A. 0.1% (m/v) NaCl   **B.** 0.9 % (m/v) NaCl   **C.** 5% (m/v) NaCl
    D. 10% (m/v) glucose   **E.** 15% (m/v) glucose

37. A solution contains 0.50 mole of $CaCl_2$ in 1000 g of water. The freezing point of the solution is
    _____ the freezing point of water.
    A. higher than   **B.** lower than   **C.** the same as

Answers to the Practice Test

1. B	**2.** W	**3.** W	**4.** B	**5.** C
6. C	**7.** E	**8.** S	**9.** I	**10.** I
11. I	**12.** C	**13.** D	**14.** E	**15.** E
16. A	**17.** C	**18.** A	**19.** D	**20.** A
21. E	**22.** C	**23.** A	**24.** E	**25.** D
26. B	**27.** D	**28.** A	**29.** C	**30.** C
31. B	**32.** A	**33.** C	**34.** C	**35.** B
36. B	**37.** B			

Acids and Bases

Study Goals

- Describe the characteristics of Arrhenius and Brønsted–Lowry acids and bases.
- Identify Brønsted–Lowry conjugate acid–base pairs.
- Use the ion product constant of water to calculate $[H_3O^+]$, $[OH^-]$, and pH.
- Write balanced equations for reactions of an acid with metals, carbonates, or bases.
- Calculate the concentration of an acid solution from titration data.
- Describe the function of a buffer.

Think About It

1. Why do a lemon, grapefruit, and vinegar taste sour?

2. What do antacids do? What are some bases listed on the labels of antacids?

3. Why are some aspirin products buffered?

Key Terms

    a. acid   **b.** base   **c.** pH   **d.** neutralization   **e.** buffer

1. ____ a substance that forms hydroxide ions (OH^-) in water and/or accepts hydrogen ions (H^+)

2. ____ a reaction between an acid and a base to form a salt and water

3. ____ a substance that forms hydrogen ions (H^+) in water

4. ____ a mixture of a weak acid (or base) and its salt that maintains the pH of a solution

5. ____ a measure of the acidity of a solution

Answers   **1.** b   **2.** d   **3.** a   **4.** e   **5.** c

8.1 Acids and Bases

- In water, an Arrhenius acid produces H_3O^+, and an Arrhenius base produces OH^-.
- The names of inorganic acids are based on the anion present.
- Hydrogen ions (H^+) from acids bond to polar water molecules in solution to form hydronium ions, H_3O^+.
- According to the Brønsted–Lowry theory, acids are hydrogen ion (H^+) donors and bases are hydrogen ion acceptors.
- Conjugate acid–base pairs are molecules or ions linked by the loss and gain of a hydrogen ion.

(MC)

Tutorial: Naming Acids and Bases
Tutorial: Acid and Base Formulas
Self Study Activity: Nature of Acids and Bases
Tutorial: Identifying Conjugate Acids–Base Pairs
Tutorial: Definitions of Acids and Bases
Tutorial: Properties of Acids and Bases

◆ **Learning Exercise 8.1A**

Indicate if the following characteristics describe an (A) acid or (B) base.

1. ____ turns blue litmus red   2. ____ tastes bitter

3. ____ tastes sour   4. ____ turns red litmus blue

5. ____ has more OH^- ions than H_3O^+ ions   6. ____ has more H_3O^+ ions than OH^- ions

7. ____ neutralizes bases   8. ____ neutralizes acids

Answers   1. A   2. B   3. A   4. B   5. B   6. A   7. A   8. B

◆ **Learning Exercise 8.1B**

Fill in the blanks with the formula or name of an acid or base:

1. HCl   _____

2. ____   sodium hydroxide

3. ____   sulfurous acid

4. ____   nitric acid

5. $Ca(OH)_2$ _____

6. H_2CO_3 _____

7. $Al(OH)_3$ _____

8. ____   potassium hydroxide

9. $HClO_3$ _____

10. H_3PO_3 _____

Answers
1. hydrochloric acid
2. NaOH
3. H_2SO_3
4. HNO_3
5. calcium hydroxide
6. carbonic acid
7. aluminum hydroxide
8. KOH
9. chloric acid
10. phosphorous acid

Study Note

Identify the conjugate acid–base pairs in the following reaction:

$$HCl + H_2O \rightarrow H_3O^+ + Cl^-$$

Solution: HCl (hydrogen-ion donor) and Cl^- (hydrogen-ion acceptor)
H_2O (hydrogen-ion acceptor) and H_3O^+ (hydrogen-ion donor)

◆ **Learning Exercise 8.1C**

Complete the following table:

Conjugate Acid	Conjugate Base
1. H_2O	_____
2. HSO_4^-	_____
3. _____	F^-
4. _____	CO_3^{2-}
5. HNO_3	_____
6. NH_4^+	_____
7. _____	HS^-
8. _____	$H_2PO_4^-$

Answers
1. OH^-
2. SO_4^{2-}
3. HF
4. HCO_3^-
5. NO_3^-
6. NH_3
7. H_2S
8. H_3PO_4

◆ **Learning Exercise 8.1D**

Identify the conjugate acid–base pairs in each of the following equations:

1. $HF(aq) + H_2O(l) \rightleftarrows H_3O^+(aq) + F^-(aq)$ _____

2. $NH_4^+(aq) + SO_4^{2-}(aq) \rightleftarrows NH_3(aq) + HSO_4^-(aq)$ _____

3. $NH_3(aq) + H_2O(l) \rightleftarrows NH_4^+(aq) + OH^-(aq)$ _____

4. $HNO_3(aq) + OH^-(aq) \rightleftarrows H_2O(l) + NO_3^-(aq)$ _____

Answers
1. HF/F^- and H_2O/H_3O^+
2. NH_4^+/NH_3 and HSO_4^-/SO_4^{2-}
3. NH_4^+/NH_3 and H_2O/OH^-
4. HNO_3/NO_3^- and H_2O/OH^-

8.2 Strengths of Acids and Bases

- In aqueous solution, a strong acid donates all of its H^+ to water, whereas a weak acid donates only a small percentage of H^+ to water.
- Most hydroxides of Groups 1A (1) and 2A (2) are strong bases, which dissociate nearly completely in water.
- In an aqueous ammonia solution, NH_3, which is a weak base, accepts only a small percentage of hydrogen ions to form NH_4^+.

Study Note

There are only six common strong acids: other acids are weak acids.

HCl	HNO_3
HBr	H_2SO_4 (first H)
HI	$HClO_4$

Example: Is H_2S a strong or weak acid?

Solution: H_2S is a weak acid because it is not one of the six strong acids.

♦ **Learning Exercise 8.2A**

Identify each of the following as a strong or weak acid or base:

1. HNO_3 _____   **2.** H_2CO_3 _____   **3.** H_3PO_4 _____

4. NH_3 _____   **5.** LiOH _____   **6.** H_3BO_3 _____

7. $Ca(OH)_2$ _____   **8.** H_2SO_4 _____

Answers **1.** strong acid   **2.** weak acid   **3.** weak acid   **4.** weak base
5. strong base   **6.** weak acid   **7.** strong base   **8.** strong acid

♦ **Learning Exercise 8.2B**

Using Table 8.3 in your textbook, identify the stronger acid in each of the following pairs of acids:

1. HCl or H_2CO_3 _____   **2.** HNO_2 or HCN _____

3. H_2S or HBr _____   **4.** H_2SO_4 or HSO_4^- _____

5. HF or H_3PO_4 _____

Answers   **1.** HCl   **2.** HNO_2 **3.** HBr   **4.** H_2SO_4 **5.** H_3PO_4

8.3 Ionization of Water

- In pure water, a few water molecules transfer H^+ to other water molecules, forming two conjugate acid–base pairs.
- The transfer of H^+ to water produces small, but equal, amounts of $[H_3O^+]$ and $[OH^-] = 1 \times 10^{-7}$ moles/L.
- K_w, the ion product constant of water, $[H_3O^+][OH^-] = [1 \times 10^{-7}][1 \times 10^{-7}] = 1 \times 10^{-14}$, applies to all aqueous solutions at 25 °C.
- In acidic solutions, the $[H_3O^+]$ is greater than the $[OH^-]$. In basic solutions, the $[OH^-]$ is greater than the $[H_3O^+]$.

Tutorial: Ionization of Water

Guide to Calculating $[H_3O^+]$ and $[OH^-]$ in Aqueous Solutions	
Step 1	Write the K_w for water.
Step 2	Solve the K_w for the unknown $[H_3O^+]$ or $[OH^-]$.
Step 3	Substitute the known $[H_3O^+]$ or $[OH^-]$ and calculate.

Study Note

Example: What is the $[H_3O^+]$ in a solution with $[OH^-] = 2.0 \times 10^{-9}$ M?

Solution: **Step 1: Write the K_w for water.**

$$K_w = [H_3O^+][OH^-] = 1.0 \times 10^{-14}$$

Step 2: Solve the K_w for the unknown $[H_3O^+]$.

$$[H_3O^+] = \frac{K_w}{[OH^-]}$$

Step 3: Substitute the known $[OH^-]$ and calculate.

$$[H_3O] = \frac{1.0 \times 10^{-14}}{2.0 \times 10^{-9}} = 5.0 \times 10^{-6} \text{ M}$$

◆ Learning Exercise 8.3A

Calculate the $[H_3O^+]$ when the $[OH^-]$ has each of the following values:

1. $[OH^-] = 1.0 \times 10^{-10}$ M $[H_3O^+] =$ _____

2. $[OH^-] = 2.0 \times 10^{-5}$ M $[H_3O^+] =$ _____

3. $[OH^-] = 4.5 \times 10^{-7}$ M $[H_3O^+] =$ _____

4. $[OH^-] = 8.0 \times 10^{-4}$ M $[H_3O^+] =$ _____

5. $[OH^-] = 5.5 \times 10^{-8}$ M $[H_3O^+] =$ _____

Answers   **1.** 1.0×10^{-4} M   **2.** 5.0×10^{-10} M   **3.** 2.2×10^{-8} M
              4. 1.3×10^{-11} M   **5.** 1.8×10^{-7} M

♦ **Learning Exercise 8.3B**

Calculate the $[OH^-]$ when the $[H_3O^+]$ has each of the following values:

1. $[H_3O^+] = 1.0 \times 10^{-3}$ M $[OH^-] = $ _____

2. $[H_3O^+] = 3.0 \times 10^{-10}$ M $[OH^-] = $ _____

3. $[H_3O^+] = 4.0 \times 10^{-6}$ M $[OH^-] = $ _____

4. $[H_3O^+] = 2.8 \times 10^{-13}$ M $[OH^-] = $ _____

5. $[H_3O^+] = 8.6 \times 10^{-7}$ M $[OH^-] = $ _____

Answers   **1.** 1.0×10^{-11} M   **2.** 3.3×10^{-5} M   **3.** 2.5×10^{-9} M
              4. 3.6×10^{-2} M   **5.** 1.2×10^{-8} M

8.4 The pH Scale

- The pH scale is a range of numbers from 0 to 14 related to the $[H_3O^+]$ of the solution.
- A neutral solution has a pH of 7. In acidic solutions, the pH is below 7.0, and in basic solutions the pH is above 7.0.
- Mathematically, pH is the negative logarithm of the hydronium ion concentration:

$$pH = -\log [H_3O^+]$$

Tutorial: The pH Scale
Case Study: Hyperventilation and Blood pH

♦ **Learning Exercise 8.4A**

State whether the following pH values are acidic, basic or neutral:

1. ____ blood plasma, pH = 7.4   **2.** ____ soft drink, pH = 2.8

3. ____ maple syrup, pH = 6.8   **4.** ____ beans, pH = 5.0

5. ____ tomatoes, pH = 4.2   **6.** ____ lemon juice, pH = 2.2

7. ____ saliva, pH = 7.0   **8.** ____ eggs, pH = 7.8

9. ____ lime, pH = 12.4   **10.** ____ strawberries, pH = 3.0

Answers   **1.** basic   **2.** acidic   **3.** acidic   **4.** acidic   **5.** acidic
              6. acidic   **7.** neutral   **8.** basic   **9.** basic   **10.** acidic

Tutorial: Logarithms	

Guide to Calculating pH of an Aqueous Solution	
Step 1	Enter the $[H_3O^+]$.
Step 2	Press the *log key* and change the sign.
Step 3	Adjust the number of SFs on the *right* of the decimal point to equal the SFs in the coefficient.

Study Note

Example: What is the pH of a solution with $[H_3O^+] = 5.0 \times 10^{-4}$ M?

Solution: **Step 1: Enter the $[H_3O^+]$.**

$$pH = -\log[H_3O^+] \quad pH = -\log[5.0 \times 10^{-4}]$$

Step 2: Press the *log key* and change the sign.

$$pH = -\log[5.0 \times 10^{-4}] = 3.3010 \quad \text{(calculator display)}$$

Step 3: Adjust the number of SFs on the *right* of the decimal point to equal the SFs in the coefficient.

$$pH = 3.30$$

◆ **Learning Exercise 8.4B**

Calculate the pH of each of the following solutions.

1. $[H_3O^+] = 1 \times 10^{-8}$ M _____

2. $[OH^-] = 1 \times 10^{-12}$ M _____

3. $[H_3O^+] = 1 \times 10^{-3}$ M _____

4. $[OH^-] = 1 \times 10^{-10}$ M _____

Answers   **1.** 8.0   **2.** 2.0   **3.** 3.0   **4.** 4.0

◆ **Learning Exercise 8.4C**

Calculate the pH of each of the following solutions:

1. $[H_3O^+] = 5.0 \times 10^{-3}$ M

2. $[OH^-] = 4.0 \times 10^{-6}$ M

3. $[H_3O^+] = 7.5 \times 10^{-8}$ M

4. $[OH^-] = 2.5 \times 10^{-10}$ M

Answers   **1.** 2.30   **2.** 8.60
          3. 7.12   **4.** 4.40

♦ **Learning Exercise 8.4D**

Complete the following table:

	$[H_3O^+]$	$[OH^-]$	pH
1.		1×10^{-12} M	
2.			8.0
3.	1×10^{-10} M		
4.			7.0
5.			1.0

Answers		$[H_3O^+]$	$[OH^-]$	pH
	1.	1×10^{-2} M	1×10^{-12} M	2.0
	2.	1×10^{-8} M	1×10^{-6} M	8.0
	3.	1×10^{-10} M	1×10^{-4} M	10.0
	4.	1×10^{-7} M	1×10^{-7} M	7.0
	5.	1×10^{-1} M	1×10^{-13} M	1.0

8.5 Reactions of Acids and Bases

- Acids react with many metals to yield hydrogen gas (H_2) and the salt of the metal.
- Acids react with carbonates and bicarbonates to yield CO_2 gas, H_2O, and the salt of the metal.
- Acids neutralize bases in a reaction that produces water and a salt.
- The net ionic equation for any neutralization of an acid and a strong base is: $H^+ + OH^- \rightarrow H_2O$.
- In a balanced neutralization equation, an equal number of moles of H^+ and OH^- must react.
- The concentration of an acid can be determined by titration.

(MC)

Self Study Activity: Nature of Acids and Bases
Tutorial: Acid-Base Titrations

♦ **Learning Exercise 8.5A**

Complete and balance each of the following reactions of acids:

1. _____ Zn(s) + _____ HCl(aq) → _____ ZnCl₂(aq) + _____

2. _____ Li₂CO₃(s) + _____ HCl(aq) → _____ + _____ + _____

3. _____ NaHCO₃(s) + _____ HCl(aq) → _____ CO₂(g) _____ + H₂O(l) + _____ NaCl(aq)

4. _____ Al(s) + _____ H₂SO₄(aq) → _____ Al₂(SO₄)₃(aq) + _____

Answers
1. Zn(s) + 2HCl(aq) → ZnCl₂(aq) + H₂(g)
2. Li₂CO₃(s) + 2HCl(aq) → CO₂(g) + H₂O(l) + 2LiCl(aq)
3. NaHCO₃(s) + HCl(aq) → CO₂(g) + H₂O(l) + NaCl(aq)
4. 2Al(s) + 3H₂SO₄(aq) → Al₂(SO₄)₃(aq) + 3H₂(g)

Guide to Balancing an Equation for Neutralization	
Step 1	Write the reactants and products.
Step 2	Balance the H^+ in the acid with the OH^- in the base.
Step 3	Balance the H_2O with the H^+ and the OH^-.
Step 4	Write the salt from the remaining ions.

♦ **Learning Exercise 8.5B**

Balance each of the following neutralization reactions:

1. $NaOH(aq) + H_2SO_4(aq) \rightarrow Na_2SO_4(aq) + H_2O(l)$

2. $Mg(OH)_2(aq) + HCl(aq) \rightarrow MgCl_2(aq) + H_2O(l)$

3. $Al(OH)_3(aq) + HNO_3(aq) \rightarrow Al(NO_3)_3(aq) + H_2O(l)$

4. $Ca(OH)_2(aq) + H_3PO_4(aq) \rightarrow Ca_3(PO_4)_2(s) + H_2O(l)$

Answers
    1. $2NaOH(aq) + H_2SO_4(aq) \rightarrow Na_2SO_4(aq) + 2H_2O(l)$
    2. $Mg(OH)_2(aq) + 2HCl(aq) \rightarrow MgCl_2(aq) + 2H_2O(l)$
    3. $Al(OH)_3(aq) + 3HNO_3(aq) \rightarrow Al(NO_3)_3(aq) + 3H_2O(l)$
    4. $3Ca(OH)_2(aq) + 2H_3PO_4(aq) \rightarrow Ca_3(PO_4)_2(s) + 6H_2O(l)$

♦ **Learning Exercise 8.5C**

Complete each of the following neutralization reactions and then balance:

1. $KOH(aq) + H_3PO_4(aq) \rightarrow$ _____ $+ H_2O(l)$

2. $NaOH(aq) +$ _____ $\rightarrow Na_2SO_4(aq) +$ _____

3. _____ $+$ _____ $\rightarrow AlCl_3(aq) +$ _____

4. _____ $+$ _____ $\rightarrow Fe_2(SO_4)_3(s) +$ _____

Answers
    1. $3KOH(aq) + H_3PO_4(aq) \rightarrow K_3PO_4(aq) + 3H_2O(l)$
    2. $2NaOH(aq) + H_2SO_4(aq) \rightarrow Na_2SO_4(aq) + 2H_2O(l)$
    3. $Al(OH)_3(aq) + 3HCl(aq) \rightarrow AlCl_3(aq) + 3H_2O(l)$
    4. $2Fe(OH)_3(aq) + 3H_2SO_4(aq) \rightarrow Fe_2(SO_4)_3(s) + 6H_2O(l)$

(MC)

Tutorial: Acid–Base Titrations

Guide to Calculations for an Acid–Base Titration	
Step 1	State the given and needed quantities.
Step 2	Write a plan to calculate molarity or volume.
Step 3	State equalities and conversion factors including concentrations.
Step 4	Set up the problem to calculate the needed quantity.

♦ **Learning Exercise 8.5D**

1. A 24.6-mL sample of HCl solution reacts with 33.0 mL of 0.222 M NaOH solution. What is the molarity of the HCl solution?

$$HCl(aq) + NaOH(aq) \rightarrow NaCl(aq) + H_2O(l)$$

2. A 15.7-mL sample of H_2SO_4 solution reacts with 27.7 mL of 0.187 M KOH solution. What is the molarity of the H_2SO_4 solution?

$$H_2SO_4(aq) + 2KOH(aq) \rightarrow K_2SO_4(aq) + 2H_2O(l)$$

3. A 10.0-mL sample of H_3PO_4 solution is placed in a flask. If titration requires 42.0 mL of a 0.100 M NaOH solution to reach the endpoint, what is the molarity of the H_3PO_4 solution?

$$H_3PO_4(aq) + 3NaOH(aq) \rightarrow Na_3PO_4(aq) + 3H_2O(l)$$

Answers 1. 0.298 M HCl solution 2. 0.165 M H_2SO_4 solution
3. 0.140 M H_3PO_4 solution

8.6 Buffers

• A buffer solution resists a change in pH when small amounts of acid or base are added.
• A buffer contains either (1) a weak acid and its salt, or (2) a weak base and its salt. In (1), the weak acid picks up excess OH^-, and the anion of the salt picks up excess H_3O^+. In (2), the weak base reacts with excess H_3O^+, and the cation of the salt reacts with excess OH^-.

♦ **Learning Exercise 8.6**

State whether each of the following represents a buffer system or not. Explain your reason.

1. HCl + NaCl _____

2. K_2SO_4 _____

3. H_2CO_3 _____

4. $H_2CO_3 + NaHCO_3$ _____

Answers 1. No. A strong acid is not a buffer.
2. No. A salt alone cannot act as a buffer.
3. No. A weak acid alone cannot act as a buffer.
4. Yes. A weak acid and its salt act as a buffer system.

Tutorial: Buffer Solutions
Tutorial: pH and Buffers

Checklist for Chapter 8

You are ready to take the practice test for Chapter 8. Be sure that you have accomplished the following learning goals for this chapter. If you are not sure, review the section listed at the end of the goal. Then apply your new skills and understanding to the practice test.

After studying Chapter 8, I can successfully:

____ Describe the properties of Arrhenius acids and bases and write their names (8.1).

____ Describe the Brønsted–Lowry concept of acids and bases; write conjugate acid–base pairs for an acid–base reaction (8.1).

____ Write equations for the ionization of strong and weak acids and bases (8.2).

____ From a given pair of acids or bases, determine which is the stronger acid or base (8.2).

____ Use the ion product constant of water to calculate $[H_3O^+]$ and $[OH^-]$ (8.3).

____ Calculate pH from the $[H_3O^+]$ of a solution (8.4).

____ Calculate $[H_3O^+]$ from the pH of a solution (8.4).

____ Write a balanced equation for the reactions of acids with metals, carbonates, or bases (8.5).

____ Calculate the molarity or volume of an acid from titration data (8.5).

____ Describe the role of buffers in maintaining the pH of a solution and calculate the pH of a buffer solution (8.6).

Practice Test for Chapter 8

1. An acid is a compound which, when placed in water, yields this characteristic ion:
   A. H_3O^+ **B.** OH^- **C.** Na^+
   D. Cl^- **E.** CO_3^{2-}

2. $MgCl_2$ would be classified as a(n)
   A. acid   **B.** base   **C.** salt
   D. buffer   **E.** nonelectrolyte

3. $Mg(OH)_2$ would be classified as a
   A. weak acid   **B.** strong base   **C.** salt
   D. buffer   **E.** nonelectrolyte

4. In the K_w expression for pure H_2O, the $[H_3O^+]$ has the value (at 25 °C) of
   A. 1×10^{-7} M   **B.** 1×10^{-1} M   **C.** 1×10^{-14} M
   D. 1×10^{-6} M   **E.** 1×10^{-12} M

5. Which of the following pH values is most acidic?
   A. 8.0   **B.** 5.5   **C.** 1.5
   D. 3.2   **E.** 9.0

6. Which of the following pH values is most basic?
   A. 10.0   B. 4.0   C. 2.2
   D. 11.5   E. 9.0

For questions 7 through 9, consider a solution with $[H_3O^+] = 1 \times 10^{-11}$ M.

7. The pH of the solution is
   A. 1.0   B. 2.0   C. 3.0
   D. 11.0   E. 14.0

8. The hydroxide ion concentration is
   A. 1×10^{-1} M   B. 1×10^{-3} M   C. 1×10^{-4} M
   D. 1×10^{-7} M   E. 1×10^{-11} M

9. The solution is
   A. acidic   B. basic   C. neutral
   D. a buffer   E. neutralized

For questions 10 through 12, consider a solution with a $[OH^-] = 1 \times 10^{-5}$ M.

10. The hydrogen ion concentration of the solution is
    A. 1×10^{-5} M   B. 1×10^{-7} M   C. 1×10^{-9} M
    D. 1×10^{-10} M   E. 1×10^{-14} M

11. The pH of the solution is
    A. 2.0   B. 5.0   C. 9.0
    D. 11   E. 14

12. The solution is
    A. acidic   B. basic   C. neutral
    D. a buffer   E. neutralized

13. Acetic acid is a weak acid because
    A. it forms a dilute acid solution.   B. it is isotonic.
    C. it is less than 50% ionized in water.   D. it is a nonpolar molecule.
    E. it can form a buffer.

14. A weak base when added to water
    A. makes the solution slightly basic.   B. does not affect the pH.
    C. dissociates completely.   D. does not dissociate.
    E. makes the solution slightly acidic.

15. Which is an equation for neutralization?
    A. $CaCO_3(s) \rightarrow CaO(s) + CO_2(g)$ B. $Na_2SO_4(s) \rightarrow 2Na^+(aq) + SO_4^{2-}(aq)$
    C. $H_2SO_4(aq) + 2NaOH(aq) \rightarrow Na_2SO_4(aq) + 2H_2O(l)$ D. $Na_2O(s) + SO_3(g) \rightarrow Na_2SO_4(s)$
    E. $H_2CO_3(s) \rightarrow CO_2(g) + H_2O(l)$

16. What is the name given to components in the body that keep blood pH within its normal 7.35 to 7.45 range?
    A. nutrients   B. buffers   C. metabolites
    D. fluids   E. neutralizers

17. What is true of a typical buffer system?
    A. It maintains a pH of 7.0.   **B.** It contains a weak base.
    C. It contains a salt.   **D.** It contains a strong acid and its salt.
    E. It maintains the pH of a solution.

18. Which of the following would act as a buffer system?
    A. HCl   **B.** Na_2CO_3 **C.** $NaOH + NaNO_3$
    D. NH_4OH **E.** $NaHCO_3 + H_2CO_3$

19. Which of the following is a conjugate acid–base pair?
    A. HCl/HNO_3 **B.** HNO_3 / NO_3^- **C.** NaOH/KOH
    D. HSO_4^- / HCO_3^- **E.** Cl^-/F^-

20. The conjugate base of HSO_4^- is
    A. SO_4^{2-} **B.** H_2SO_4 **C.** HS^-
    D. H_2S **E.** SO_4^{2-}

21. In which reaction does H_2O act as an acid?
    A. $H_3PO_4(aq) + H_2O(l) \rightarrow H_3O^+(aq) + H_2PO_4^-(aq)$
    B. $H_2SO_4(aq) + H_2O(l) \rightarrow H_3O^+(aq) + HSO_4^-(aq)$
    C. $H_2O(l) + HS^-(aq) \rightarrow H_3O^+(aq) + S^{2-}(aq)$
    D. $NaOH(aq) + HCl(aq) \rightarrow NaCl(aq) + H_2O(l)$
    E. $NH_3(aq) + H_2O(l) \rightarrow NH_4^+(aq) + OH^-(aq)$

22. If 23.7 mL of HCl solution reacts with 19.6 mL of 0.179 M NaOH solution, what is the molarity of the HCl solution?
    A. 6.76 M   **B.** 0.216 M   **C.** 0.148 M
    D. 0.163 M   **E.** 0.333 M

23. A solution has a pH of 4.0. It has a $[H_3O^+]$ of
    A. 1×10^{-4} M   **B.** 1×10^4 M   **C.** 1×10^{-10} M
    D. 4.0 M   **E.** 1×10^{-14} M

24. A solution has a pH of 8.0. It has an $[OH^-]$ of
    A. 1×10^{-8} M   **B.** 1×10^8 M   **C.** 1×10^{-6} M
    D. 6.0 M   **E.** 1×10^{-14} M

25. A solution has a pH of 5.7. It has a $[H_3O^+]$ of
    A. 5×10^{-9} M   **B.** 2×10^{-6} M   **C.** 5×10^{-7} M
    D. 2×10^{-5} M   **E.** 1×10^{-14} M

Answers to the Practice Test

1. A	**2.** C	**3.** B	**4.** A	**5.** C
6. D	**7.** D	**8.** B	**9.** B	**10.** C
11. C	**12.** B	**13.** C	**14.** A	**15.** C
16. B	**17.** E	**18.** E	**19.** B	**20.** A
21. E	**22.** C	**23.** A	**24.** C	**25.** B

Nuclear Radiation

Study Goals

- Describe alpha, beta, positron, and gamma radiation.
- Describe the methods required for proper shielding for each type of radiation.
- Write an equation showing the mass numbers and atomic numbers for an atom that undergoes radioactive decay or bombardment.
- Describe the detection and measurement of radiation.
- Describe the use of radioisotopes in nuclear medicine.
- Calculate the amount of radioisotope that remains after a given number of half-lives.
- Describe nuclear fission and fusion.

Think About It

1. What is nuclear radiation?

2. Why do you receive more radiation if you live in the mountains or travel on an airplane?

3. In nuclear medicine, iodine-125 is used for detecting a tumor in the thyroid. What does the number 125 indicate?

4. Why is there a concern about radon in our homes?

5. How does nuclear fission differ from nuclear fusion?

Key Terms

Match each of the following key terms with the correct definition:

a. radiation	**b.** half-life	**c.** curie	**d.** positron
e. nuclear fission	**f.** alpha particle	**g.** rem	

1. _____ a particle identical to a helium nucleus produced in a radioactive nucleus

2. _____ the time required for one-half of a radioactive sample to undergo radioactive decay

3. _____ a unit of radiation measurement equal to 3.7×10^{10} disintegrations per second

4. _____ a measure of biological damage caused by radiation

5. _____ a process in which large nuclei split into smaller nuclei with the release of energy

6. _____ energy or particles released by radioactive atoms

7. _____ produced when a proton is transformed into a neutron

Answers   1. f   2. b   3. c   4. g   5. e   6. a   7. d

9.1 Natural Radioactivity

- Radioactive isotopes have unstable nuclei that break down (decay), spontaneously emitting alpha (α), beta (β), positron (β^+), and gamma (γ) radiation.
- An alpha particle is the same as a helium nucleus; it contains 2 protons and 2 neutrons.
- A beta particle is a high-energy electron and a positron is a high-energy positive particle. A gamma ray is very high-energy radiation.
- Because radiation can damage cells in the body, proper protection must be used: shielding, time limitation, and distance.

(MC)

Self Study Activity: Nuclear Chemistry
Tutorial: Radiation and Its Biological Effects
Tutorial: Types of Radiation

Study Note

It is important to learn the symbols for the radiation particles in order to describe the different types of radiation:

$_1^1H$ or p	$_0^1n$ or n	$_{-1}^0e$ or β	$_2^4He$ or α	$_{+1}^0e$ or β^+	$_0^0\gamma$ or γ
proton	neutron	beta particle	alpha particle	positron	gamma ray

---

Learning Check 9.1A

Match the description in column B with the terms in column A:

A	B
1. _____ $_8^{18}O$	**a.** symbol for a beta particle
2. _____ γ	**b.** symbol for an alpha particle
3. _____ β^+	**c.** an atom that emits radiation
4. _____ radioactive isotope	**d.** symbol of a positron
5. _____ $_2^4He$	**e.** symbol for an atom of oxygen
6. _____ β	**f.** symbol for gamma radiation

133

Answers   **1.** e   **2.** f   **3.** d   **4.** c   **5.** b   **6.** a

Learning Check 9.1B

Discuss some things you can do to minimize the amount of radiation received if you work with
a radioactive substance. Describe how each method helps to limit the amount of radiation you would
receive.

Answer

Three ways to minimize exposure to radiation are: (1) use shielding, (2) keep time short in the radioactive
area, and (3) keep as much distance as possible from the radioactive materials. Shielding such as clothing
and gloves stops alpha and beta particles from reaching your skin, whereas lead or concrete will absorb
gamma rays. Limiting the time spent near radioactive samples reduces exposure time. Increasing the
distance from a radioactive source reduces the intensity of radiation. Wearing a film badge will monitor the
amount of radiation you receive.

Learning Check 9.1C

What type(s) of radiation (alpha, beta, and/or gamma) would each of the following shielding materials
protect you from?

1. clothing _____   **2.** skin _____

3. paper _____   **4.** concrete _____

5. lead wall _____

Answers   **1.** alpha, beta   **2.** alpha   **3.** alpha
            4. alpha, beta, gamma   **5.** alpha, beta, gamma

9.2 Nuclear Equations

- A balanced nuclear equation represents the changes in the nuclei of radioisotopes.
- The new isotope and the type of radiation emitted can be determined from the symbols that show the
  mass numbers and atomic numbers of the isotopes in the nuclear reaction.

Radioactive nucleus → New nucleus + Radiation

Total of the mass numbers is equal

$$^{11}_{6}C \rightarrow {}^{7}_{4}Be + {}^{4}_{2}He$$

Total of the atomic numbers is equal

- A new radioactive isotope is produced when a nonradioactive isotope is bombarded by a small
  particle such as a proton, an alpha particle, or a beta particle.

$$\underset{\substack{\text{Bombarding} \\ \text{particle } (\alpha)}}{^{4}_{2}He} \quad + \quad \underset{\substack{\text{Stable} \\ \text{nucleus}}}{^{10}_{5}B} \quad \rightarrow \quad \underset{\substack{\text{New} \\ \text{nucleus}}}{^{13}_{7}N} \quad + \quad \underset{\substack{\text{Emitted} \\ \text{neutron}}}{^{1}_{0}n}$$

Guide to Completing a Nuclear Equation	
Step 1	Write the incomplete nuclear equation.
Step 2	Determine the missing mass number.
Step 3	Determine the missing atomic number.
Step 4	Determine the symbol of the new nucleus.
Step 5	Complete the nuclear equation.

(MC)

Tutorial: Writing Nuclear Equations
Tutorial: Alpha, Beta, and Gamma Emitters

Study Note

When balancing nuclear equations for radioactive decay, be sure that
1. The mass number of the reactant is equal to the sum of the mass numbers of the products.
2. The atomic number of the reactant is equal to the sum of the atomic numbers of the products.

Changes in Mass Number and Atomic Number Due to Radiation

Decay Process	Radiation Symbol	Change in Mass Number	Change in Atomic Number	Change in Neutron Number
Alpha emission	$^{4}_{2}He$	−4	−2	−2
Beta emission	$^{0}_{-1}e$	0	+1	−1
Positron emission	$^{0}_{+1}e$	0	−1	+1
Gamma emission	$^{0}_{0}\gamma$	0	0	0

Learning Check 9.2A

Write a nuclear symbol that completes each of the following nuclear equations:

1. $^{66}_{29}Cu \rightarrow {}^{66}_{30}Zn + ?$ 1. _____

2. $^{127}_{53}I \rightarrow {}^{1}_{0}n + ?$ 2. _____

3. $^{238}_{92}U \rightarrow {}^{4}_{2}He + ?$ 3. _____

4. $^{24}_{11}Na \rightarrow {}^{0}_{-1}e + ?$ 4. _____

5. $? \rightarrow {}^{30}_{14}Si + {}^{0}_{-1}e$ 5. _____

Answers   1. $^{0}_{-1}e$ 2. $^{126}_{53}I$ 3. $^{234}_{90}Th$ 4. $^{24}_{12}Mg$ 5. $^{30}_{13}Al$

Study Note

In balancing nuclear transmutation equations (bombardment by small particles),

1. The sum of the mass numbers of the reactants must equal the sum of the mass numbers of the products.
2. The sum of the atomic numbers of the reactants must equal the sum of the atomic numbers of the products.

Learning Check 9.2B

Complete each of the following equations for bombardment reactions:

a. $? + {}^{40}_{20}Ca \rightarrow {}^{40}_{19}K + {}^{1}_{1}H$ a. _____

b. ${}^{1}_{0}n + {}^{27}_{13}Al \rightarrow {}^{24}_{11}Na + ?$ b. _____

c. ${}^{1}_{0}n + {}^{10}_{5}B \rightarrow {}^{4}_{2}He + ?$ c. _____

d. $? + {}^{23}_{11}Na \rightarrow {}^{23}_{12}Mg + {}^{1}_{0}n$ d. _____

e. ${}^{1}_{1}H + {}^{197}_{79}Au \rightarrow ? + {}^{1}_{0}n$ e. _____

Answers   a. ${}^{1}_{0}n$ b. ${}^{4}_{2}He$ c. ${}^{7}_{3}Li$ d. ${}^{1}_{1}H$ e. ${}^{197}_{80}Hg$

9.3 Radiation Measurement

- A Geiger counter is used to detect radiation. When radiation passes through the gas in the counter tube, some atoms of gas are ionized, producing an electrical current.
- The activity of a radioactive sample measures the number of nuclear transformations per second. The curie (Ci) is equal to 3.7×10^{10} disintegrations per second. The becquerel (Bq) is equal to 1 disintegration per second.
- The radiation dose absorbed by a gram of body tissue is measured in units of rads or the SI unit grays, which is equal to 100 rad.
- The biological damage of different types of radiation on the body is measured in radiation units of rems or sieverts (Sv), which is equal to 100 rem.

MC

Tutorial: Measuring Radiation
Case Study: Food Irradiation

Learning Check 9.3A

Match each type of measurement unit with the radiation process measured.

    a. curie   **b.** becquerel   **c.** rad   **d.** gray   **e.** rem

1. _____ an activity of one disintegration per second

2. _____ the amount of radiation absorbed by 1 g of material

3. _____ an activity of 3.7×10^{10} disintegrations per second

4. _____ the biological damage caused by different kinds of radiation

5. _____ a unit of absorbed dose equal to 100 rad

Answers   **1.** b   **2.** c   **3.** a   **4.** e   **5.** d

Learning Check 9.3B

1. A sample of Ir-192 has an activity of 35 μCi. What is its activity in becquerels?

2. If an absorbed dose is 15 mrem, what is the absorbed dose in sieverts?

Answers   **1.** 1.3×10^{6} Bq   **2.** 1.5×10^{-4} Sv

9.4 Half-Life of a Radioisotope

- The half-life of a radioactive sample is the time required for one-half of the sample to decay (emit radiation).
- Most radioisotopes used in medicine, such as Tc-99m and I-131, have short half-lives. By comparison, many naturally occurring radioisotopes, such as C-14, Ra-226, and U-238, have long half-lives. For example, potassium-42 has a half-life of 12 h, whereas potassium-40 takes 1.3×10^{9} y for one-half of the radioactive sample to decay.

Tutorial: Radioactive Half-Lives
Tutorial: Radiocarbon Dating

Guide to Using Half-Lives	
Step 1	State the given and needed quantities.
Step 2	Write a plan to calculate the unknown quantity.
Step 3	Write the half-life equality and conversion factors.
Step 4	Set up the problem to calculate amount of active radioisotope.

Study Note

In one half-life, one-half of the initial quantity of a radioisotope emits radiation, while one-half of the sample remains active.

Example: Chromium-51 has a half-life of 28 days. How much of a 16-μg sample of chromium-51 remains after 84 days?

Solution:

Step 1: **State the given and needed quantities.**

Given 16-μg sample of Cr-51; 84 days; 28 days/1 half-life

Need μg of Cr-51 remaining

Step 2: **Write a plan to calculate the unknown quantity.**

84 days $\xrightarrow{\text{half-life}}$ Number of half-lives

16-μg of Cr-51 $\xrightarrow{\text{Number of half-lives}}$ μg of Cr-51 remaining

Step 3: **Write the half-life equality and conversion factors.**

1 half-life = 28 days

$$\frac{28 \text{ days}}{1 \text{ half-life}} \quad \text{and} \quad \frac{1 \text{ half-life}}{28 \text{ days}}$$

Step 4: **Set up the problem to calculate amount of active radioisotope.**

$$\text{Number of half-lives} = 84 \text{ days} \times \frac{1 \text{ half-life}}{28 \text{ days}} = 3 \text{ half-lives}$$

Now, we can calculate the quantity of Cr-51 that remains active

$$16 \ \mu g \xrightarrow{1 \text{ half-life}} 8.0 \ \mu g \xrightarrow{1 \text{ half-life}} 4.0 \ \mu g \xrightarrow{1 \text{ half-life}} 2.0 \ \mu g$$

Learning Check 9.4

1. Suppose you have an 80-mg sample of iodine-125. If iodine-125 has a half-life of 60 days, how many mg are radioactive
   a. after one half-life?

   b. after two half-lives?

   c. after 240 days?

2. $^{99m}_{43}$Tc has a half-life of 6 h. A technician picked up a 16-mg sample at 8 A.M.; how much of the sample remained radioactive at 8 P.M. that same day?

3. Flourine-18 has a half-life of 110 minutes. How much of a 240-μg sample will be radioactive after 440 minutes?

4. Iodine-131 has a half-life of 8.0 days. How many days will it take for 80 mg of I-131 to decay to 5 mg?

5. Suppose a group of archaeologists digs up some pieces of a wooden boat at an ancient site. When a sample of the wood is analyzed for C-14, scientists determine that 12.5%, or 1/8, of the original amount of C-14 remains. If the half-life of carbon-14 is 5730 years, how long ago was the boat made?

Answers **1. a.** 40. mg   **b.** 20. mg   **c.** 5.0 mg   **2.** 4.0 mg
3. 15 μg   **4.** 32 days   **5.** 17 200 y ago

9.5 Medical Applications Using Radioactivity

- Nuclear medicine uses radioactive isotopes that go to specific sites in the body.
- For diagnostic work, radioisotopes are used that emit gamma rays and produce nonradioactive products.
- By detecting the radiation emitted by medical radioisotopes, evaluations can be made about the location and extent of an injury, disease, or tumor, blood flow, or level of function of a particular organ.

Learning Check 9.5

Write the nuclear symbol for each of the following radioactive isotopes:

1. _____ iodine-131 used to study thyroid gland activity

2. _____ phosphorus-32 used to locate brain tumors

3. _____ sodium-24 used to determine blood flow and to locate a blood clot or embolism

4. _____ nitrogen-13 used in positron emission tomography

Answers   **1.** $^{131}_{53}\text{I}$ **2.** $^{32}_{15}\text{P}$ **3.** $^{24}_{11}\text{Na}$ **4.** $^{13}_{7}\text{N}$

9.6 Nuclear Fission and Fusion

- In fission, a large nucleus breaks apart into smaller pieces, releasing one or more types of radiation and a great amount of energy.
- A chain reaction is a fission reaction that will continue once started.
- In fusion, small nuclei combine to form a larger nucleus, which releases great amounts of energy.
- Nuclear fission is currently used to produce electrical energy while nuclear fusion is still in the experimental stage.

Tutorial: Fission and Fusion
Tutorial: Nuclear Fission and Fusion Reactions

Learning Check 9.6A

Discuss the nuclear processes of fission and fusion for the production of energy.

Answer

Nuclear fission is a splitting of the atom into two or more nuclei accompanied by the release of large amounts of energy and radiation. In the process of *nuclear fusion*, two or more nuclei combine to form a heavier nucleus and release a large amount of energy. However, fusion requires a considerable amount of energy to initiate the process.

Learning Check 9.6B

Balance each of the following nuclear equations and identify each as a fission or fusion reaction:

1. $_0^1n + _{92}^{235}U \longrightarrow _{54}^{143}Xe + 3_0^1n + ?$

2. $_1^1H + _1^2H \longrightarrow ?$

3. $? + _1^3H \longrightarrow _2^4He + _0^1n$

4. $_0^1n + _{92}^{235}U \longrightarrow _{36}^{91}Kr + 3_0^1n + ?$

Answers   1. $_{38}^{90}Sr$, fission   2. $_2^3He$, fusion   3. $_1^2H$, fusion   4. $_{56}^{142}Ba$, fission

Checklist for Chapter 9

You are ready to take the Practice Test for Chapter 9. Be sure you have accomplished the following learning goals for this chapter. If you are not sure, review the section listed at the end of the goal. Then apply your new skills and understanding to the Practice Test.

After studying Chapter 9, I can successfully:

_____ Describe alpha, beta, and gamma radiation (9.1).

_____ Write a nuclear equation showing mass numbers and atomic numbers for radioactive

decay (9.2).

_____ Write a nuclear equation for the formation of a radioactive isotope (9.2).

_____ Describe the detection and measurement of radiation (9.3).

_____ Calculate the amount of a radioisotope remaining after one or more half-lives (9.4).

_____ Describe the use of radioisotopes in medicine (9.5).

_____ Describe the processes of nuclear fission and fusion (9.6).

Practice Test for Chapter 9

1. The correctly written symbol for an isotope of sulfur is
   A. $^{30}_{16}$Su   B. $^{14}_{30}$Si   C. $^{30}_{16}$S   D. $^{30}_{16}$Si   E. $^{16}_{30}$S

2. Alpha particles are composed of
   A. protons   B. neutrons   C. electrons
   D. protons and electrons   E. protons and neutrons

3. Gamma radiation is a type of radiation that
   A. originates in the electron shells
   B. is most dangerous
   C. is least dangerous
   D. is the heaviest
   E. travels the shortest distance

4. The charge on an alpha particle is
   A. -1 B. $+1$ C. -2 D. $+2$ E. $+4$

5. Beta particles formed in a radioactive nucleus are
   A. protons   B. neutrons   C. electrons
   D. protons and electrons   E. protons and neutrons

For questions 6 through 10, select from the following:
   A. $^{0}_{-1}$X   B. $^{4}_{2}$X   C. $^{1}_{1}$X   D. $^{1}_{0}$X   E. $^{0}_{0}$X

6. an alpha particle

7. a beta particle

8. a gamma ray

9. a proton

10. a neutron

11. Shielding from gamma rays is provided by
    A. skin.   B. paper.   C. clothing.   D. lead.   E. air.

12. The skin will provide shielding from
    A. alpha particles   B. beta particles   C. gamma rays
    D. ultraviolet rays   E. X rays

13. The radioisotope iodine-131 is used as a radioactive tracer for studying thyroid gland activity. The symbol for iodine-131 is
    A. I   B. $_{131}$I   C. $^{131}_{53}$I   D. $^{53}_{131}$I   E. $^{78}_{53}$I

14. When an atom emits an alpha particle, its mass number will
    A. increase by 1   B. increase by 2   C. increase by 4
    D. decrease by 4   E. not change

15. When a nucleus emits a positron, the atomic number of the new nucleus
   A. increases by 1   **B.** increases by 2   **C.** decreases by 1
   D. decreases by 2   **E.** will not change

16. When a nucleus emits a gamma ray, the atomic number of the new nucleus
   A. increases by 1   **B.** increases by 2   **C.** decreases by 1
   D. decreases by 2   **E.** will not change

For questions 17 through 20, select the particle that completes each of the equations.
   A. neutron   **B.** alpha particle   **C.** beta particle   **D.** gamma ray

17. $^{126}_{50}Sn \rightarrow ^{126}_{51}Sb + ?$

18. $^{13}_{4}Be \rightarrow ^{12}_{4}Be + ?$

19. $^{99m}_{43}Tc \rightarrow ^{99}_{43}Tc + ?$

20. $^{149}_{62}Sm \rightarrow ^{145}_{60}Nd + ?$

21. What symbol completes the following reaction?

$$^{1}_{0}n + ^{14}_{7}N \rightarrow ? + ^{1}_{1}H$$

   A. $^{15}_{8}O$ **B.** $^{15}_{6}C$ **C.** $^{14}_{8}O$ **D.** $^{14}_{6}C$ **E.** $^{15}_{7}N$

22. To complete this nuclear equation, you need to write

$$? + ^{54}_{26}Fe \rightarrow ^{57}_{28}Ni + ^{1}_{0}n$$

   A. an alpha particle   **B.** a beta particle   **C.** a gamma ray
   D. a neutron   **E.** a proton

23. The name of the unit used to measure the number of disintegrations per second is
   A. curie   **B.** rad   **C.** rem   **D.** gray   **E.** sievert

24. The rem and the sievert are units used to measure
   A. activity of a radioactive sample
   B. biological damage of different types of radiation
   C. radiation absorbed
   D. background radiation
   E. half-life of a radioactive sample

25. Radiation can cause
   A. nausea   **B.** a lower white cell count
   C. fatigue   **D.** hair loss
   E. all of these

26. Radioisotopes used in medical diagnosis
   A. have short half-lives   **B.** emit gamma rays, beta particles, or positrons
   C. locate in specific organs   **D.** produce nonradioactive nuclei
   E. all of these

27. The time required for a radioisotope to decay is measured by its
    A. half-life   **B.** protons   **C.** activity
    D. fusion   **E.** radioisotope

28. Oxygen-15 used in imaging has a half-life of 2 min. How many half-lives have occurred in the 10 minutes it takes to prepare the sample?
    A. 2   **B.** 3   **C.** 4   **D.** 5   **E.** 6

29. Iodine-131 has a half-life of 8 days. How long will it take for a 160-mg sample to decay to 10 mg?
    A. 8 days   **B.** 16 days   **C.** 32 days   **D.** 40 days   **E.** 48 days

30. Palladium-103 has a half-life of 17 days. After 34 days, how many milligrams of a 100-mg sample will still be radioactive?
    A. 75 mg   **B.** 50 mg   **C.** 40 mg   **D.** 25 mg   **E.** 12.5 mg

31. The "splitting" of a large nucleus to form smaller particles accompanied by a release of energy is called
    A. radioisotope   **B.** fission   **C.** fusion   **D.** rem   **E.** half-life

32. The process of combining small nuclei to form larger nuclei is
    A. radioisotope   **B.** fission   **C.** fusion   **D.** rem   **E.** half-life

33. The fusion reaction
    A. occurs in the Sun
    B. forms larger nuclei from smaller nuclei
    C. requires extremely high temperatures
    D. releases a large amount of energy
    E. all of these

Answers to the Practice Test

1. C	**2.** E	**3.** B	**4.** D	**5.** C
6. B	**7.** A	**8.** E	**9.** C	**10.** D
11. D	**12.** A	**13.** C	**14.** D	**15.** C
16. E	**17.** C	**18.** A	**19.** D	**20.** B
21. D	**22.** A	**23.** A	**24.** B	**25.** E
26. E	**27.** A	**28.** D	**29.** C	**30.** D
31. B	**32.** C	**33.** E		

10

Introduction to Organic Chemistry: Alkanes

Study Goals

- Describe the properties that are characteristic of organic compounds.
- Identify the number of bonds for carbon and other atoms in organic compounds.
- Describe the tetrahedral shape of carbon with single bonds in organic compounds.
- Draw condensed structural formulas and skeletal formulas for alkanes.
- Write the IUPAC names for alkanes and cycloalkanes.
- Write equations for the combustion of alkanes.
- Identify the functional groups in organic compounds.

Think About It

1. What is the meaning of the term "organic"?

2. What two elements are found in all organic compounds?

3. In some salad dressings, why is there a layer of vegetable oil floating on the vinegar and water layer?

Key Terms

1. Match the statements below with the following key terms:

   a. alkene   **b.** isomers   **c.** hydrocarbon
   d. alcohol   **e.** functional group   **f.** alkane
   g. condensed structural formula   **h.** main chain   **i.** combustion
   j. cycloalkane

1. _____ an atom or group of atoms that influences the chemical reactions of an organic compound

2. _____ a class of organic compounds with one or more hydroxyl (—OH) group

3. _____ a type of hydrocarbon with one or more carbon–carbon double bond

4. _____ an organic compound consisting of only carbon and hydrogen atoms

5. _____ compounds having the same molecular formula but a different arrangement of atoms

6. _____ a hydrocarbon that contains only carbon–carbon single bonds

7. _____ an alkane that exists as a cyclic structure

8. _____ the chemical reaction of an alkane and oxygen that yields CO_2, H_2O, and heat

9. _____ the type of formula that shows the arrangement of the carbon atoms grouped with their attached H atoms

10. _____ the longest continuous chain of carbon atoms in a condensed structural formula

Answers   **1.** e   **2.** d   **3.** a   **4.** c   **5.** b
                6. f   **7.** j   **8.** i   **9.** g   **10.** h

10.1 Organic Compounds

- Organic compounds are compounds of carbon and hydrogen that have covalent bonds, have low melting and boiling points, burn vigorously in air, are mostly nonpolar molecules, and are usually more soluble in nonpolar solvents than in water.
- Each carbon in an alkane has four bonds arranged so that the bonded atoms are in the corners of a tetrahedron.
- Alkanes are hydrocarbons that have only single bonds, C—C and C—H.

(MC)

Self Study Activity: Introduction to Organic Molecules

♦ **Learning Exercise 10.1A**

Identify each the following as typical of an organic (O) or inorganic (I) compound:

1. _____ has covalent bonds   2. _____ has a low boiling point

3. _____ burns in air   4. _____ is soluble in water

5. _____ has a high melting point   6. _____ is soluble in a nonpolar solvent

7. _____ has ionic bonds   8. _____ has a long carbon chain

9. _____ contains carbon   10. _____ does not burn in air

11. _____ has a formula of Na_2SO_4 12. _____ has a formula of CH_3—CH_2—CH_3

Answers   **1.** O   **2.** O   **3.** O   **4.** I   **5.** I   **6.** O
                7. I   **8.** O   **9.** O   **10.** I   **11.** I   **12.** O

♦ **Learning Exercise 10.1B**

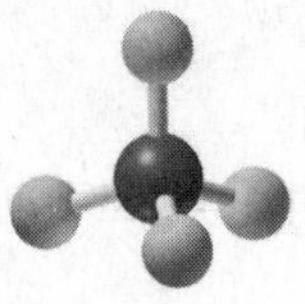

1. What is the name of the three-dimensional shape of methane shown above?

2. Draw the expanded structural formula (two-dimensional) for methane.

Answers   1.   tetrahedron

2. 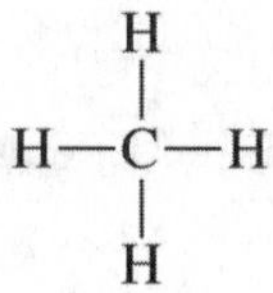 $\text{H}-\overset{\displaystyle \text{H}}{\underset{\displaystyle \text{H}}{\text{C}}}-\text{H}$

10.2 Alkanes

- An expanded structural formula shows a separate line to each bonded atom; a condensed structural formula depicts each carbon atom and its attached hydrogen atoms as a group. A skeletal formula shows the carbon atoms as the corners or ends of a zigzag line. A molecular formula gives the total number of each kind of atom.

Expanded Structural Formula	Condensed Structural Formula	Skeletal Formula	Molecular Formula
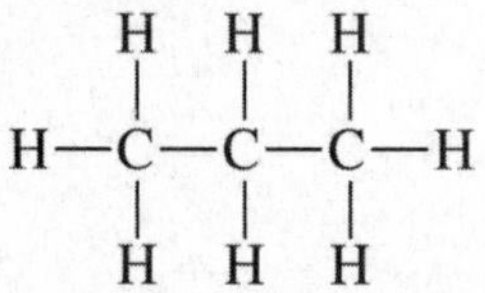 $\text{H}-\overset{\text{H}}{\underset{\text{H}}{\text{C}}}-\overset{\text{H}}{\underset{\text{H}}{\text{C}}}-\overset{\text{H}}{\underset{\text{H}}{\text{C}}}-\text{H}$	$CH_3-CH_2-CH_3$	 ⌃	C_3H_8

- In an IUPAC (International Union of Pure and Applied Chemistry) name, the prefix indicates the number of carbon atoms, and the suffix describes the family of the compound. For example, in the name *propane,* the prefix *prop* indicates a chain of three carbon atoms and the ending *ane* indicates single bonds (alkane). The names of the first six alkanes follow:

Name	Number of Carbon Atoms	Condensed Structural Formula
Methane	1	CH_4
Ethane	2	$CH_3—CH_3$
Propane	3	$CH_3—CH_2—CH_3$
Butane	4	$CH_3—CH_2—CH_2—CH_3$
Pentane	5	$CH_3—CH_2—CH_2—CH_2—CH_3$
Hexane	6	$CH_3—CH_2—CH_2—CH_2—CH_2—CH_3$

◆ **Learning Exercise 10.2A**

Indicate if each of the following is, an expanded structural formula (E), a condensed structural formula (C), a skeletal formula (S) or a molecular formula (M):

1. ____ $CH_3—CH_3$ 2. ____ C_5H_{12} 3. ____ $CH_3—CH_2—CH_3$

4.   5. 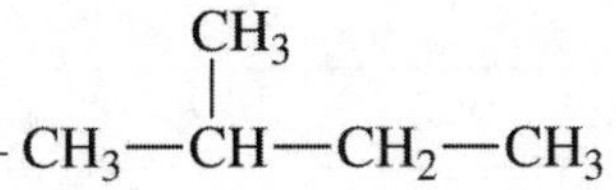   6.

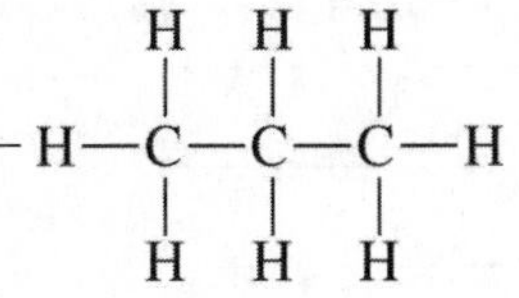

Answers   1. C   2. M   3. C   4. E   5. C   6. S

◆ **Learning Exercise 10.2B**

Draw the condensed structural formula for each of the following expanded structural formulas:

1.

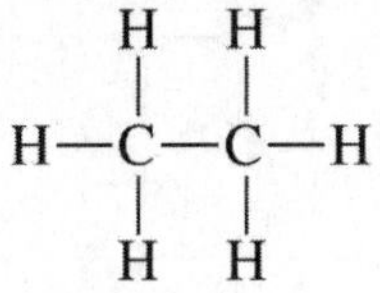

2.

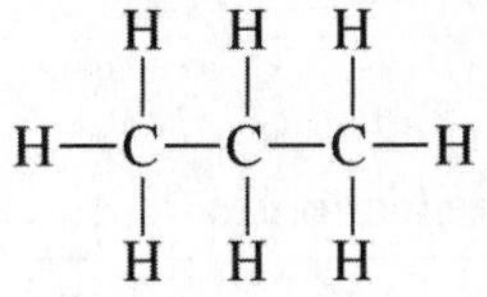

3.

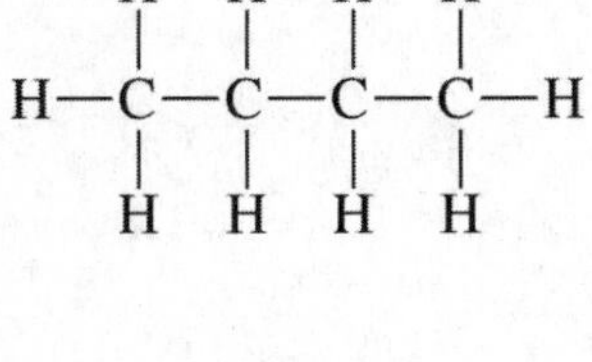

4.

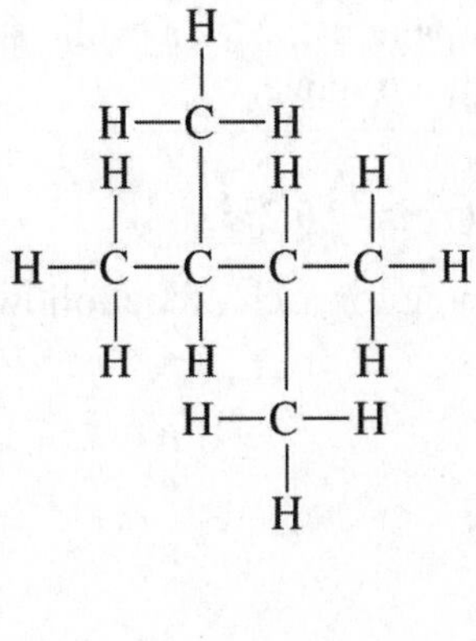

147

Answers   1. CH$_3$—CH$_3$ 2. CH$_3$—CH$_2$—CH$_3$

                                                       CH$_3$
                                                        |
3. CH$_3$—CH$_2$—CH$_2$—CH$_3$ 4. CH$_3$—CH—CH—CH$_3$
                                                        |
                                                       CH$_3$

Tutorial: IUPAC Naming of Alkanes

♦ **Learning Exercise 10.2C**

Draw the condensed structural formula and give the name for the continuous-chain alkane with each of the following molecular formulas:

1. C$_2$H$_6$ _____

2. C$_3$H$_8$ _____

3. C$_4$H$_{10}$ _____

4. C$_5$H$_{12}$ _____

5. C$_6$H$_{14}$ _____

Answers   1. CH$_3$—CH$_3$, ethane
                 2. CH$_3$—CH$_2$—CH$_3$, propane
                 3. CH$_3$—CH$_2$—CH$_2$—CH$_3$, butane
                 4. CH$_3$—CH$_2$—CH$_2$—CH$_2$—CH$_3$, pentane
                 5. CH$_3$—CH$_2$—CH$_2$—CH$_2$—CH$_2$—CH$_3$, hexane

---

Tutorial: Naming Cycloalkanes

Study Note

A cycloalkane is named by adding the prefix *cyclo* to the name of the corresponding alkane. For example, the name cyclopropane indicates a cyclic structure of three carbon atoms, usually represented by the geometric shape of a triangle.

♦ **Learning Exercise 10.2D**

Give the IUPAC name for each of the following cycloalkanes:

1. ⬡   2. ☐   3. ⬠   4. △

Answers   1. cyclohexane   2. cyclobutane   3. cyclopentane   4. cyclopropane

♦ **Learning Exercise 10.2E**

Draw the skeletal formula for each of the following:

1. butane   _____

2. pentane   _____

3. hexane   _____

4. octane   _____

5. cyclohexane   _____

Answers

1.  /\/   2.  /\/\   3.  /\/\/

4.  /\/\/\/   5. 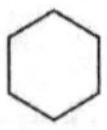 ⬡

10.3 Alkanes with Substituents

- Isomers have the same molecular formula but differ in the sequence of atoms in each of their condensed structural formulas.
- In the IUPAC system each substituent is numbered and listed alphabetically in front of the name of the longest chain.
- Carbon groups that are substituents are named as alkyl groups or alkyl substituents. An alkyl group is named by replacing the *ane* of the alkane name with *yl*. For example, CH_3— is methyl (from CH_4 methane), and CH_3—CH_2— is ethyl (from CH_3—CH_3 ethane).
- In a haloalkane, a halogen atom, —F, —Cl, —Br, or —I, replaces a hydrogen atom in an alkane.
- A halogen atom is named as a substituent (fluoro, chloro, bromo, iodo) attached to the alkane chain.

Tutorial: Naming Alkanes with Substituents

Study Note

Example: Write the IUPAC name for the following compound:

$$CH_3-CH_2-\underset{\underset{CH_3}{|}}{CH}-CH_3$$

Solution: The four-carbon chain butane is numbered from the end nearer the side group, which places the *methyl* substituent on carbon 2: 2-methylbutane.

Guide to Naming Alkanes	
Step 1	Write the alkane name of the longest chain of carbon atoms.
Step 2	Number the carbon atoms starting from the end nearer a substituent.
Step 3	Give the location and name of each substituent (alphabetical order) as a prefix to the name of the main chain.

♦ **Learning Exercise 10.3A**

Give the IUPAC name for each of the following compounds:

        CH₃
        |
1. 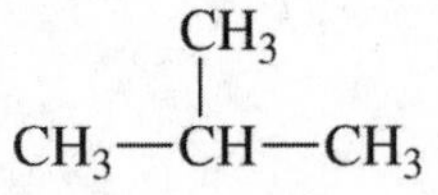 CH₃—CH—CH₃   _____

        CH₃   CH₃
        | |
2. 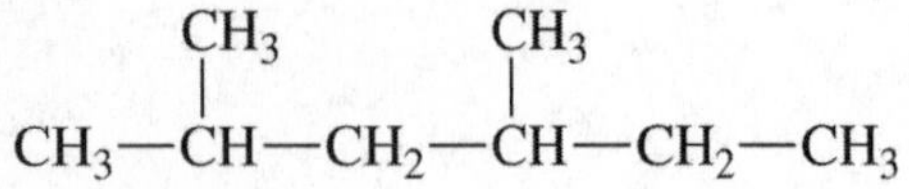 CH₃—CH—CH₂—CH—CH₂—CH₃   _____

3.   _____

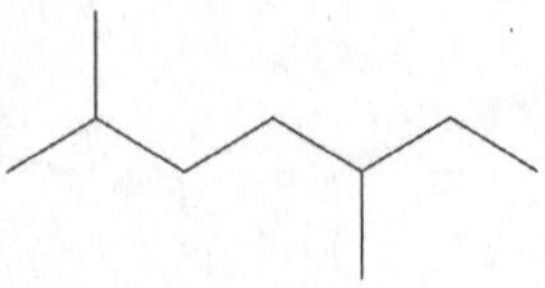

        CH₃
        |
4. 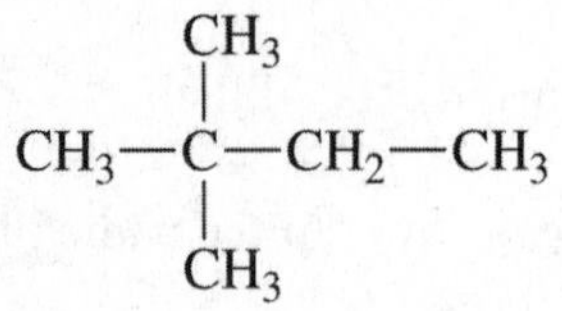 CH₃—C—CH₂—CH₃   _____
        |
        CH₃

Answers   **1.** 2-methylpropane   **2.** 2,4-dimethylhexane   **3.** 2,5-dimethylheptane
                 4. 2,2-dimethylbutane

(MC)

Tutorial: Drawing Haloalkanes and Branched Alkanes
Tutorial: Drawing Organic Compounds with Functional Groups
Self Study Activity: Organic Molecules and Isomers

Guide to Drawing Alkane Formulas	
Step 1	Draw the main chain of carbon atoms.
Step 2	Number the chain and place the substituents on the carbons indicated by the numbers.
Step 3	Add the correct number of hydrogen atoms to give four bonds to each C atom.

♦ **Learning Exercise 10.3B**

Draw the condensed structural formulas for **1.**, **2.**, and **3.** and the skeletal formulas for **4.**, **5.**, and **6.**

1. pentane

2. 2-methylpentane

3. 4-ethyl-2-methylhexane

4. 2,2,4-trimethylhexane

5. 1,2-dichlorobutane

6. methylcyclohexane

Answers

1. $CH_3—CH_2—CH_2—CH_2—CH_3$

2.
$$CH_3—\overset{\overset{\displaystyle CH_3}{|}}{CH}—CH_2—CH_2—CH_3$$

3.
$$CH_3—\overset{\overset{\displaystyle CH_3}{|}}{CH}—CH_2—\overset{\overset{\displaystyle CH_2—CH_3}{|}}{CH}—CH_2—CH_3$$

4.

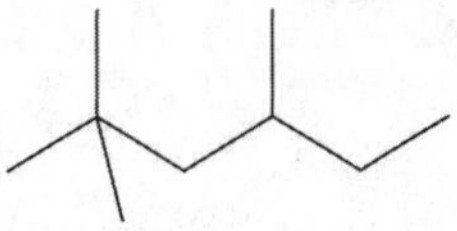

5.

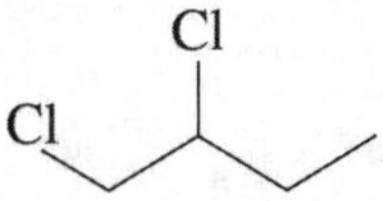

6.

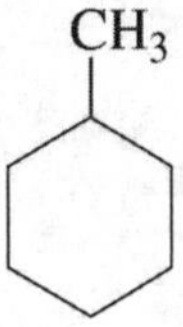

♦ **Learning Exercise 10.3C**

Give the IUPAC name for each of the following:

1. $CH_3—CH_2—Br$

2.
$$CH_3—CH_2—\overset{\overset{\displaystyle Cl}{|}}{\underset{\underset{\displaystyle Cl}{|}}{C}}—CH_2—CH_3$$

3.

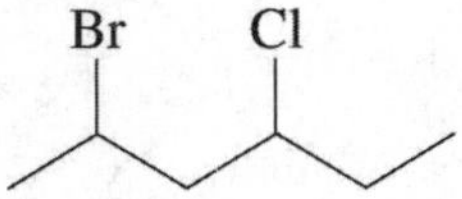

4.
$$CH_3—CH_2—CH_2—\overset{\overset{\displaystyle F}{|}}{CH}—Cl$$

5.

Answers
1. bromoethane
2. 3,3-dichloropentane
3. 2-bromo-4-chlorohexane
4. 1-chloro-1-fluorobutane
5. bromocyclopropane

♦ **Learning Exercise 10.3D**

Draw the condensed structural formulas for **1.**, **2.**, and **3.** and the skeletal formulas for **4.**, **5.**, and **6.**

1. ethyl chloride

2. bromomethane

3. 3-bromo-1-chloropentane

4. 1,1-dichlorohexane

5. 2,2,3-trichlorobutane

6. 2,4-dibromo-2,4-dichloropentane

Answers

1. CH_3-CH_2-Cl

2. CH_3-Br

3. $Cl-CH_2-CH_2-\overset{\overset{\displaystyle Br}{|}}{C}H-CH_2-CH_3$

4.

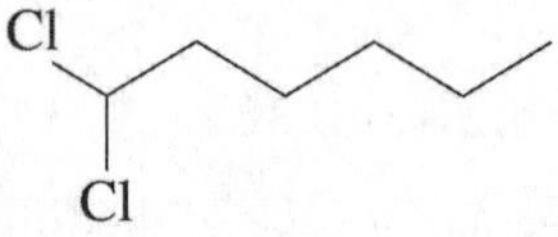

5.

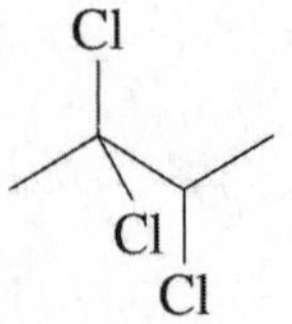

6.

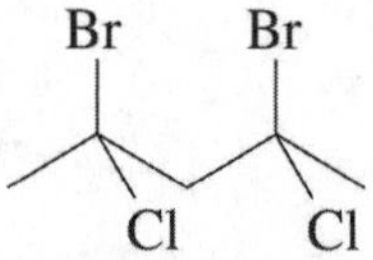

10.4 Properties of Alkanes

- Alkanes are less dense than water, and mostly unreactive, except that they burn vigorously.
- Alkanes are nonpolar and insoluble in water.
- Alkanes are found in natural gas, gasoline, and diesel fuels.
- In combustion, an alkane reacts rapidly with oxygen at a high temperature to produce carbon dioxide, water, and a great amount of heat.

Tutorial: Writing Balanced Equations for the Combustion of Alkanes
Case Study: Hazardous Materials
Case Study: Poison in the Home: Carbon Monoxide

Study Note

Example: Write the equation for the combustion of methane.
Solution: Write the molecular formulas for the following reactants: methane (CH_4) and oxygen (O_2).
Write the products CO_2 and H_2O and balance the equation.

$$CH_4(g) + O_2(g) \xrightarrow{\Delta} CO_2(g) + H_2O(g) + heat$$

$$CH_4(g) + 2O_2(g) \xrightarrow{\Delta} CO_2(g) + 2H_2O(g) + heat \text{ (balanced)}$$

♦ **Learning Exercise 10.4A**

Write a balanced equation for the complete combustion of each of the following:

1. propane _____

2. hexane _____

3. pentane _____

4. cyclobutane _____

Answers

1. $C_3H_8(g) + 5O_2(g) \xrightarrow{\Delta} 3CO_2(g) + 4H_2O(g) + heat$

2. $2C_6H_{14}(g) + 19O_2(g) \xrightarrow{\Delta} 12CO_2(g) + 14H_2O(g) + heat$

3. $C_5H_{12}(g) + 8O_2(g) \xrightarrow{\Delta} 5CO_2(g) + 6H_2O(g) + heat$

4. $C_4H_8(g) + 6O_2(g) \xrightarrow{\Delta} 4CO_2(g) + 4H_2O(g) + heat$

♦ **Learning Exercise 10.4B**

Indicate which property in each of the following pairs is more likely to be a property of hexane:

a. density is 0.66 g/mL or 1.66 g/mL _____

b. solid or liquid at room temperature _____

c. soluble or insoluble in water _____

d. flammable or nonflammable _____

e. melting point is 95 °C or −95 °C _____

Answers   **a.** 0.66 g/mL   **b.** liquid   **c.** insoluble
              d. flammable   **e.** −95 °C

10.5 Functional Groups

- Organic compounds are classified by *functional groups*, atoms or groups of atoms that determine physical and chemical properties and names.
- Alkenes are hydrocarbons that contain one or more double bonds ($C=C$); alkynes contain a triple bond ($C\equiv C$). Aromatic compounds contain a benzene ring.
- Alcohols contain a hydroxyl group (—OH); ethers have an oxygen atom (—O—) between two alkyl or aromatic groups. Thiols contain a thiol group (—SH) bonded to a carbon atom.
- Aldehydes contain a carbonyl group ($C=O$) bonded to at least one H atom; ketones contain the carbonyl group bonded to two alkyl or aromatic groups.
- Carboxylic acids have a carboxyl group (—COO) attached to hydrogen; esters contain the carboxyl group attached to an alkyl or aromatic group.
- Amines are derived from ammonia (NH_3) in which alkyl or aromatic groups replace one or more of the H atoms of NH_3.
- Amides have a carbonyl group attached to nitrogen from NH_3 or amines.

Tutorial: Identifying Functional Groups
Tutorial: Drawing Organic Compounds with Functional Groups
Self Study Activity: Functional Groups

♦ **Learning Exercise 10.5A**

Classify each of the following organic compounds **1** through **6** as:

    a. alkane   **b.** alkene   **c.** alcohol   **d.** ether   **e.** thiol

    1. ____ CH_3—CH_2—$CH=CH_2$ **2.** ____ CH_3—CH_2—CH_3

    3. ____ CH_3—CH_2—$\overset{\overset{\textstyle SH}{|}}{CH}$—$CH_3$ **4.** ____ ⟋⟍⟋ OH

    5. ____ CH_3—CH_2—O—CH_2—CH_3 **6.** ____ ⟋⟍⟋⟍

Answers   **1.** b   **2.** a   **3.** e   **4.** c   **5.** d   **6.** a

♦ **Learning Exercise 10.5B**

Classify each of the following organic compounds **1** through **8** as:

    a. aldehyde   **b.** amide   **c.** ketone   **d.** ester   **e.** amine   **f.** carboxylic acid

    1. ____ CH_3—CH_2—CH_2—$\overset{\overset{\textstyle O}{\|}}{C}$—$NH_2$ **2.** ____ CH_3—CH_2—CH_2—NH_2

    3. ____ CH_3—$\overset{\overset{\textstyle O}{\|}}{C}$—$O$—$CH_3$ **4.** ____ CH_3—CH_2—$\overset{\overset{\textstyle O}{\|}}{C}$—$OH$

5. _____ (structure: O double bonded to central carbon, CH$_3$ on one side, CH$_2$CH$_3$ on other — a ketone)

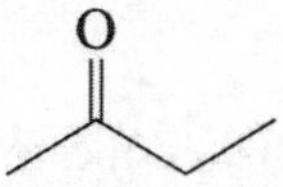

6. _____ CH$_3$—C(=O)—H

7. _____ CH$_3$—CH$_2$—CH(NH$_2$)—CH$_3$

8. _____ CH$_3$—CH$_2$—C(=O)—H

Answers

	1. b	**2.** e	**3.** d	**4.** f
	5. c	**6.** a	**7.** e	**8.** a

♦ **Learning Exercise 10.5C**

Draw the condensed structural formula for each of the following:

1. a two-carbon alcohol

2. a three-carbon ketone

3. a three-carbon aldehyde

4. a three-carbon alkene

5. an amine with two methyl groups

6. an ether with two ethyl groups

Answers

1. CH$_3$—CH$_2$—OH

2. CH$_3$—C(=O)—CH$_3$

3. CH$_3$—CH$_2$—C(=O)—H

4. CH$_3$—CH=CH$_2$

5. CH$_3$—N(H)—CH$_3$

6. CH$_3$—CH$_2$—O—CH$_2$—CH$_3$

Checklist for Chapter 10

You are ready to take the Practice Test for Chapter 10. Be sure that you have accomplished the following learning goals for this chapter. If you are not sure, review the section listed at the end of the goal. Then apply your new skills and understanding to the Practice Test.

After studying Chapter 10, I can successfully:

_____ Identify properties as characteristic of organic or inorganic compounds (10.1).

_____ Identify the number of bonds for carbon (10.1).

_____ Describe the tetrahedral shape around carbon in organic compounds (10.1).

_____ Draw the expanded structural formula and the condensed structural formula for an alkane (10.2).

_____ Use the IUPAC system to write the names for alkanes and cycloalkanes (10.2).

_____ Use the IUPAC system to write the names for alkanes with substituents (10.3).

_____ Draw the condensed structural formulas of alkanes from the name (10.3).

_____ Describe some physical properties of alkanes (10.4).

_____ Write and balance equations for the combustion of alkanes (10.4).

_____ Identify the functional groups in organic compounds (10.5).

Practice Test for Chapter 10

For questions 1 through 8, indicate whether the following characteristic are typical of organic (O) compounds or inorganic (I) compounds.

1. _____ high melting points   2. _____ fewer compounds

3. _____ covalent bonds   4. _____ soluble in water

5. _____ ionic bonds   6. _____ flammable

7. _____ low boiling points   8. _____ soluble in nonpolar solvents

For questions 9 through 13, match the formula with the correct name:
   A. methane   **B.** ethane   **C.** propane   **D.** pentane   **E.** heptane

9. _____ $CH_3-CH_2-CH_3$ 10. _____ $CH_3-CH_2-CH_2-CH_2-CH_2-CH_2-CH_3$

11. _____ CH_4 12. _____

13. _____ CH_3-CH_3

For questions 14 through 17, match the formula with the correct name:
   A. butane   **B.** 1,1-dichlorobutane   **C.** 1,2-dichloropropane
   D. 3,5-dimethylhexane   **E.** 2,4-dimethylhexane   **F.** 1,2-dichlorobutane

14. $CH_3-CH_2-CH_2-CH_3$

$$15.\ CH_3-\overset{\overset{\displaystyle CH_3}{|}}{CH}-CH_2-\overset{\overset{\displaystyle CH_3}{|}}{CH}-CH_2-CH_3$$

16.

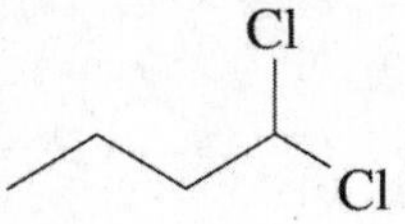

17. 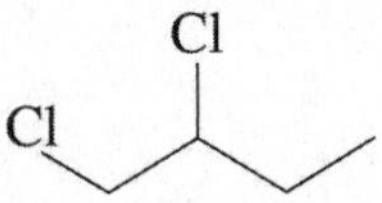 Cl

For questions 18 through 20, match the formula with the correct name:

A. methylcyclopentane
B. cyclobutane
C. cyclohexane
D. ethylcyclopentane

18. ____

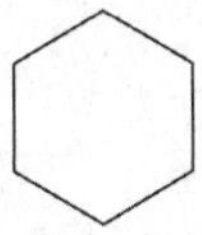

CH_3

19. ____

CH_2—CH_3

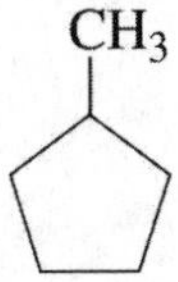

20. ____

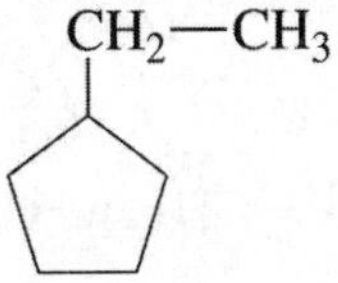

For questions 21 through 23, match the formula with the correct name:

A. 2,4-dichloropentane
B. chlorocyclopentane
C. 1,2-dichloropentane
D. 4,5-dichloropentane

21. ____ CH_3—CH_2—CH_2—$\overset{\displaystyle Cl}{\underset{|}{CH}}$—$CH_2$—$Cl$

22. ____

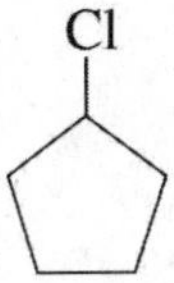

Cl

23. ____ CH_3—$\overset{\displaystyle Cl}{\underset{|}{CH}}$—$CH_2$—$\overset{\displaystyle Cl}{\underset{|}{CH}}$—$CH_3$

24. The correctly balanced equation for the complete combustion of ethane is

A. $C_2H_6(g) + O_2(g) \xrightarrow{\Delta} 2CO(g) + 3H_2O(g) +$ heat

B. $C_2H_6(g) + O_2(g) \xrightarrow{\Delta} CO_2(g) + H_2O(g) +$ heat

C. $C_2H_6(g) + 2O_2(g) \xrightarrow{\Delta} 2CO_2(g) + 3H_2O(g) +$ heat

D. $2C_2H_6(g) + 7O_2(g) \xrightarrow{\Delta} 4CO_2(g) + 6H_2O(g) +$ heat

E. $2C_2H_6(g) + 4O_2(g) \xrightarrow{\Delta} 4CO_2(g) + 6H_2O(g) +$ heat

For questions 25 through 33, match each of the formulas with its functional group:

A. thiol   **B.** alkene   **C.** alcohol   **D.** aldehyde
E. ketone   **F.** ether   **G.** amine   **H.** amide

25. ____ CH_3—CH_2—$\overset{\displaystyle O}{\overset{\displaystyle \|}{C}}$—$NH_2$

26. ____ CH_3—CH_2—$\overset{\displaystyle O}{\overset{\displaystyle \|}{C}}$—$CH_3$

27. ____
OH

28. ____ CH$_3$—CH$_2$—$\overset{\overset{\text{SH}}{|}}{\text{CH}}$—CH$_3$ 29. ____ CH$_3$—CH$_2$—O—CH$_3$

30. ____ CH$_3$—$\overset{\overset{\text{NH}_2}{|}}{\text{CH}}$—CH$_2$—CH$_3$ 31. ____ CH$_3$—$\overset{\overset{\text{O}}{||}}{\text{C}}$—H

32. ____ CH$_3$—CH$_2$—CH=CH$_2$ 33. ____ $\overset{\text{OH}}{\diagup\!\!\diagup}$

For questions 34 through 36, indicate whether the pairs of compounds are isomers (I) or the same compound (S).

34. ____ CH$_3$—CH$_2$—CH$_2$—CH$_3$ and $\overset{\overset{\text{CH}_3}{|}}{\text{CH}_2}$—CH$_2$—CH$_3$

35. ____ CH$_3$—CH$_2$—OH and CH$_3$—O—CH$_3$

36. ____ CH$_3$—CH$_2$—$\overset{\overset{\text{O}}{||}}{\text{C}}$—OH and CH$_3$—$\overset{\overset{\text{O}}{||}}{\text{C}}$—O—CH$_3$

Answers to the Practice Test

1. I	2. I	3. O	4. I	5. I
6. O	7. O	8. O	9. C	10. E
11. A	12. D	13. B	14. A	15. E
16. B	17. F	18. C	19. A	20. D
21. C	22. B	23. A	24. D	25. H
26. E	27. C	28. A	29. F	30. G
31. D	32. B	33. C	34. S	35. I
36. I				

Study Goals

- Classify unsaturated compounds as alkenes, cycloalkenes, or alkynes.
- Write IUPAC and common names for alkenes and alkynes.
- Draw condensed structural formulas and names for cis–trans isomers of alkenes.
- Write equations for the hydrogenation, halogenation, and hydration of alkenes and alkynes.
- Describe the formation of a polymer from alkene monomers.
- Describe the bonding in benzene.
- Draw condensed structural formulas and give the names of aromatic compounds.

Think About It

1. The label on a bottle of vegetable oil says the oil is unsaturated. What does this mean?

2. A margarine is partially hydrogenated. What does that mean?

3. What are polymers?

Key Terms

Match the statements shown below with the following key terms:

    a. alkene   **b.** hydrogenation   **c.** alkyne   **d.** hydration   **e.** polymer

1. _____ a long-chain molecule formed by linking many small molecules
2. _____ the addition of H_2 to a carbon–carbon double bond
3. _____ an unsaturated hydrocarbon containing a carbon–carbon double bond
4. _____ the addition of H_2O to a carbon–carbon double bond
5. _____ a compound that contains a carbon–carbon triple bond

Answers   **1.** e   **2.** b   **3.** a   **4.** d   **5.** c

11.1 Alkenes and Alkynes

- Alkenes are unsaturated hydrocarbons that contain one or more carbon–carbon double bonds.
- In alkenes, the three groups bonded to each carbon in the double bond are planar and arranged at angles of 120°.
- Alkynes are unsaturated hydrocarbons that contain a carbon–carbon triple bond.
- The IUPAC names of alkenes are derived by changing the *ane* ending of the parent alkane to *ene*. For example, the IUPAC name of $H_2C=CH_2$ is ethene. It has a common name of ethylene. In alkenes, the longest carbon chain containing the double bond is numbered from the end nearer the double bond. In cycloalkenes with substituents, the double bond carbons are given positions of 1 and 2, and the ring is numbered to give the next lower numbers to the substituents.

$$CH_3-CH=CH_2 \qquad CH_2=CH-CH_2-CH_3 \qquad CH_3-CH=CH-CH_3$$

    Propene (propylene)   1-Butene   2-Butene

- The IUPAC names of alkynes are derived by changing the *ane* ending of the parent alkane to *yne*. In alkynes, the longest carbon chain containing the triple bond is numbered from the end nearer the triple bond.

$$HC\equiv CH \qquad\qquad CH_3-C\equiv CH \qquad\qquad CH_3-CH_2-C\equiv CH$$

     Ethyne   Propyne   1-Butyne

♦ Learning Exercise 11.1A

Classify each of the following as alkane, alkene, cycloalkene, or alkyne:

1. _____ $CH_3-CH_2-CH_3$

2. _____ (structure)

3. _____ $CH_3-C\equiv C-CH_3$

4. _____ $CH_3-CH_2-CH=\overset{\displaystyle CH_3}{\overset{|}{C}}-CH_2-CH_3$

Answers   1. alkane   2. alkene   3. alkyne   4. alkene

MC

Tutorial: Naming Alkenes and Alkynes

Guide to Naming Alkenes and Alkynes
Step 1	Name the longest carbon chain that contains the double or triple bond.
Step 2	Number the carbon chain starting from the end nearer the double or triple bond.
Step 3	Give the location and name of each substituents (alphabetical order) as a prefix to the alkene or alkyne name.

♦ **Learning Exercise 11.1B**

Write the IUPAC name for each of the following alkenes:

1. $CH_3—CH{=}CH_2$

2. $CH_3—\overset{\overset{\displaystyle CH_3}{|}}{C}{=}CH—CH_3$

3. 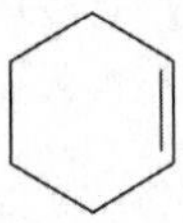 (cyclohexene structure)

4. $CH_2{=}CH—\overset{\overset{\displaystyle Cl}{|}}{CH}—CH_2—\overset{\overset{\displaystyle CH_3}{|}}{CH}—CH_3$

5. $CH_3—CH{=}\overset{\overset{\displaystyle CH_3}{|}}{C}—CH_2—CH_3$

6.

Answers   1. propene   **2.** 2-methyl-2-butene   **3.** cyclohexene
            4. 3-chloro-5-methyl-1-hexene   **5.** 3-methyl-2-pentene   **6.** 1-butene

♦ **Learning Exercise 11.1C**

Write the IUPAC name of each of the following alkynes:

1. $HC{\equiv}CH$

2. $CH_3—C{\equiv}CH$

3. $CH_3—CH_2—C{\equiv}CH$

4. $CH_3—\overset{\overset{\displaystyle CH_3}{|}}{CH}—C{\equiv}C—CH_3$

Answers   1. ethyne   2. propyne
            3. 1-butyne   4. 4-methyl-2-pentyne

Tutorial: Naming Alkenes and Alkynes
Tutorial: Drawing Alkenes and Alkynes

♦ **Learning Exercise 11.1D**

Draw the condensed structural formulas for **1.** and **2.** and the skeletal formulas for **3.** and **4.**

1. 2-pentyne

2. 2-chloro-2-butene

3. 3-bromo-2-methyl-2-pentene

4. 3-methylcyclohexene

Answers

1. CH_3—$C{\equiv}C$—CH_2—CH_3

2. CH_3—$\overset{\overset{\displaystyle Cl}{|}}{C}$=$CH$—$CH_3$

3.

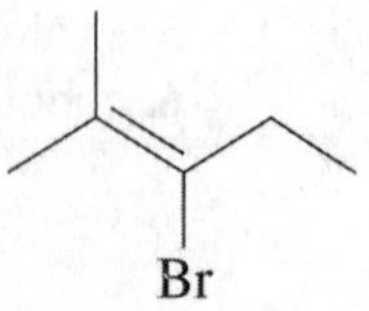

4.

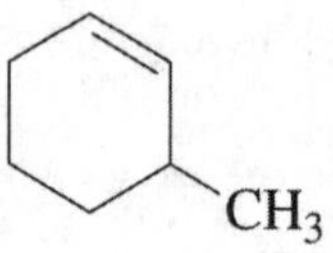

11.2 Cis–Trans Isomers

- Cis–trans isomers are possible for alkenes because there is no rotation around the rigid double bond.
- In the cis isomer, hydrogen atoms are attached on the same side of the double bond, whereas in the trans isomer, they are attached on the opposite sides of the double bond.

Tutorial: Cis–Trans Isomers

♦ **Learning Exercise 11.2A**

Draw the condensed structural formulas of the cis and trans isomers of 1,2-dibromoethene.

Answers In the cis isomer, the hydrogen atoms are attached on the same side of the double bond; in the trans isomer, they are on opposite sides.

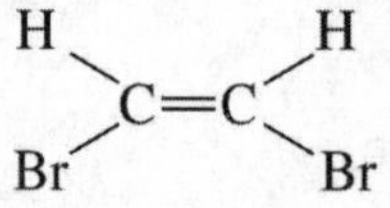

cis-1,2-Dibromoethene 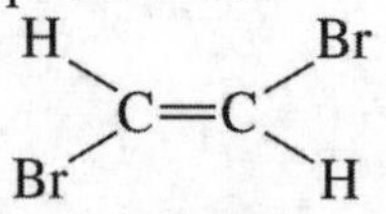   *trans*-1,2-Dibromoethene

♦ Learning Exercise 11.2B

Name the following alkenes using the cis–trans prefix where isomers are possible:

1. 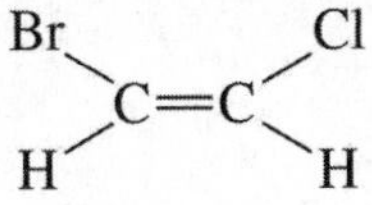 Br, Cl
C=C
H, H

_____

2. 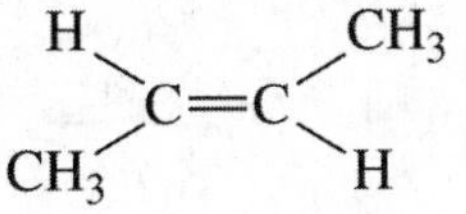 H, CH₃
C=C
CH₃, H

_____

3. 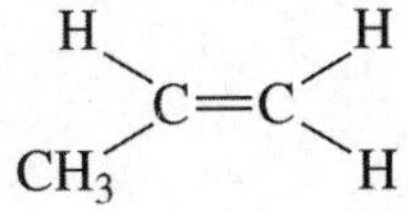 H, H
C=C
CH₃, H

_____

4. 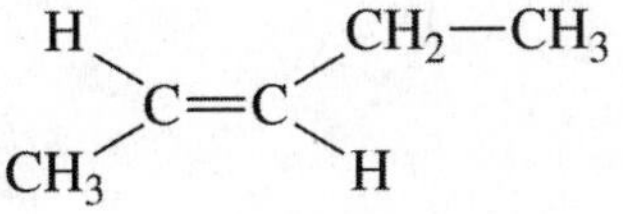 H, CH₂—CH₃
C=C
CH₃, H

_____

Answers
1. *cis*-1-bromo-2-chloroethene
3. propene (not a cis–trans isomer)

2. *trans*-2-butene
4. *trans*-2-pentene

11.3 Addition Reactions

- The addition of small molecules to the double bond is a characteristic reaction of alkenes.
- Hydrogenation adds hydrogen atoms to the double bond of an alkene or the triple bond of an alkyne to yield an alkane.

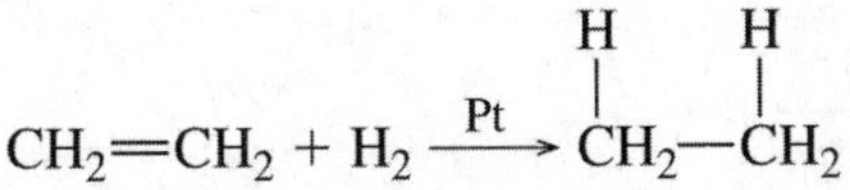

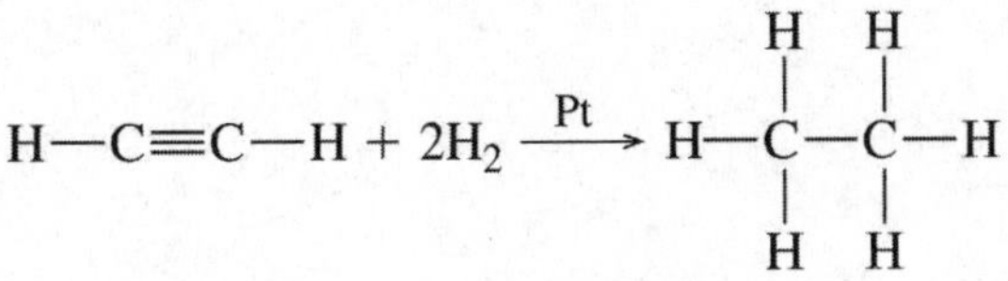

$$CH_2{=}CH_2 + H_2 \xrightarrow{Pt} \overset{\displaystyle H}{\underset{}{C}}H_2{-}\overset{\displaystyle H}{\underset{}{C}}H_2$$

$$H{-}C{\equiv}C{-}H + 2H_2 \xrightarrow{Pt} H{-}\overset{H}{\underset{H}{C}}{-}\overset{H}{\underset{H}{C}}{-}H$$

- Halogenation adds bromine or chlorine atoms to produce dihaloalkanes.

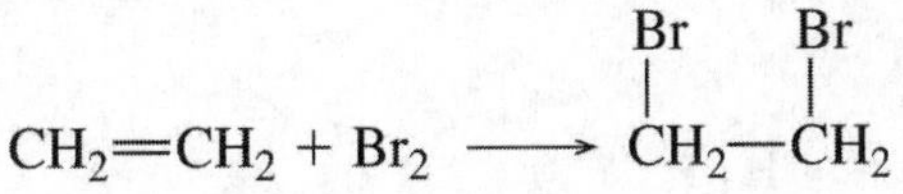

$$CH_2{=}CH_2 + Br_2 \longrightarrow \overset{Br}{\underset{}{C}}H_2{-}\overset{Br}{\underset{}{C}}H_2$$

- Hydration, in the presence of a strong acid, adds water to a double bond. The H from the reactant HOH bonds to the carbon in the double bond that has the greater number of hydrogen atoms.

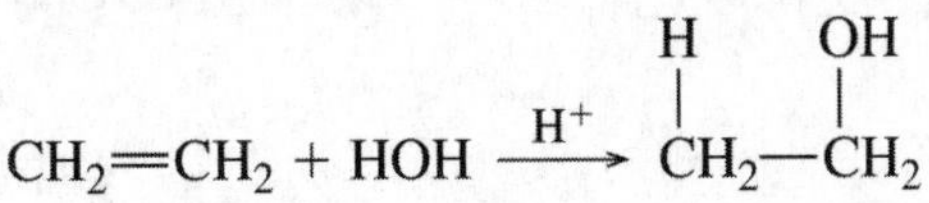

$$CH_2{=}CH_2 + HOH \xrightarrow{H^+} \overset{H}{\underset{}{C}}H_2{-}\overset{OH}{\underset{}{C}}H_2$$

Tutorial: Addition Reactions

♦ **Learning Exercise 11.3**

Draw the condensed structural formulas or skeletal formulas of the products of the following addition reactions:

1. $CH_3-CH_2-CH=CH_2 + H_2 \xrightarrow{Pt}$

2. 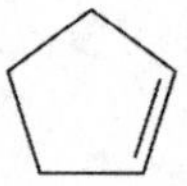 $+ H_2 \xrightarrow{Pt}$

3. $CH_3-CH=CH-CH_2-CH_3 + Cl_2 \longrightarrow$

4. $CH_3-C\equiv CH + 2H_2 \xrightarrow{Pt}$

5.  $+ Br_2 \longrightarrow$

6. $CH_3-CH=CH_2 + HOH \xrightarrow{H^+}$

7. 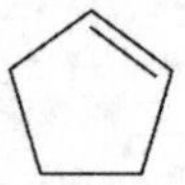 $+ HOH \xrightarrow{H^+}$

Answers

1. $CH_3-CH_2-CH_2-CH_3$

2.

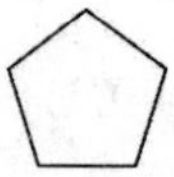

3. $CH_3-\underset{\underset{Cl}{|}}{CH}-\underset{\underset{Cl}{|}}{CH}-CH_2-CH_3$

4. $CH_3-CH_2-CH_3$

5.

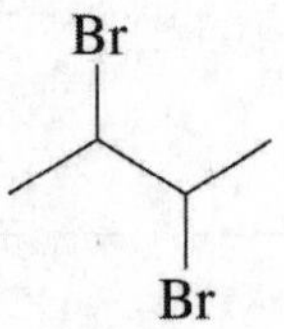

6. $CH_3-\underset{\underset{OH}{|}}{CH}-CH_3$

7.

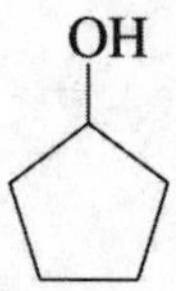

11.4 Polymers of Alkenes

- Polymers are large molecules prepared from the bonding of many small units called *monomers*.
- Many synthetic polymers are made from small alkene monomers.

(MC)

Tutorial: Polymers

♦ **Learning Check 11.4A**

Draw the condensed structural formula of the starting monomer for each of the following polymers:

1. 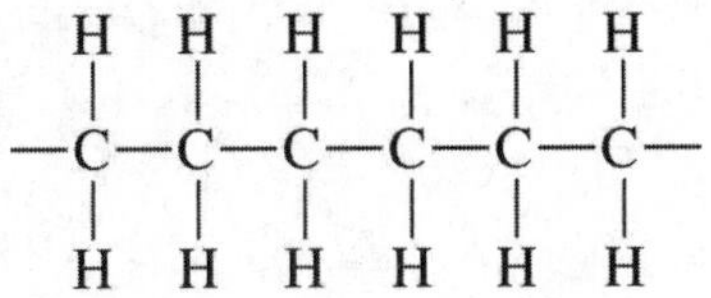
```
     H   H   H   H   H   H
     |   |   |   |   |   |
  —C—C—C—C—C—C—
     |   |   |   |   |   |
     H   H   H   H   H   H
```

2. 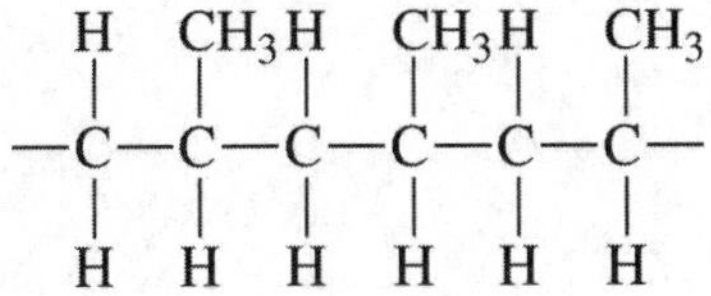
```
     H  CH₃ H  CH₃ H   CH₃
     |   |   |   |   |   |
  —C—C—C—C—C—C—
     |   |   |   |   |   |
     H   H   H   H   H   H
```

3. 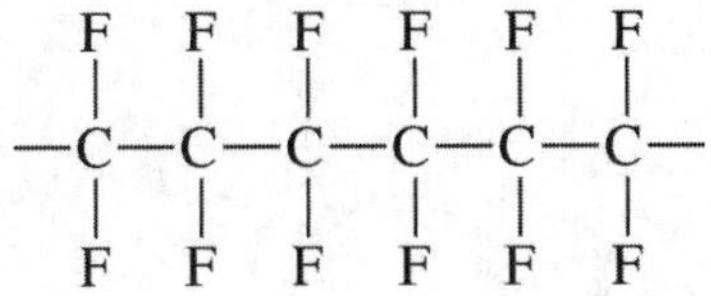
```
     F   F   F   F   F   F
     |   |   |   |   |   |
  —C—C—C—C—C—C—
     |   |   |   |   |   |
     F   F   F   F   F   F
```

Answers

1. $H_2C{=}CH_2$
2. $H_2C{=}\overset{\displaystyle CH_3}{CH}$
3. $F_2C{=}CF_2$

♦ **Learning Check 11.4B**

Draw the structure of the polymer formed from the addition of three monomers of 1,1-difluoroethene.

Answer
```
     F   H   F   H   F   H
     |   |   |   |   |   |
  —C—C—C—C—C—C—
     |   |   |   |   |   |
     F   H   F   H   F   H
```

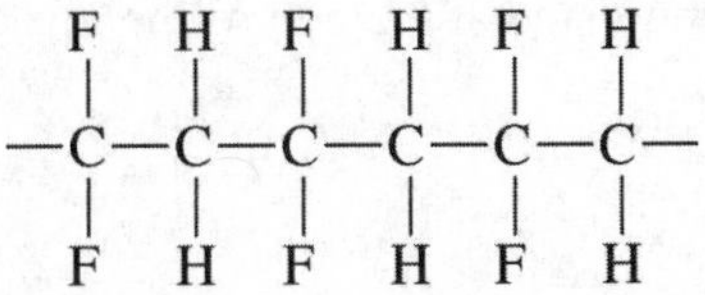

11.5 Aromatic Compounds

- Most aromatic compounds contain a benzene ring, a cyclic structure containing six CH units. The structure of benzene is represented as a hexagon with a circle in the center.
- The names of many aromatic compounds use the parent name benzene, although many common names were retained as IUPAC names, such as toluene, phenol, and aniline. The position of two substituents on the ring is shown as 1,2-, 1,3-, or 1,4-.

(MC)

Tutorial: Naming Aromatic Compounds

♦ **Learning Exercise 11.5A**

Write the IUPAC (and common) name for each of the following:

1.

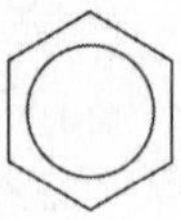

2.

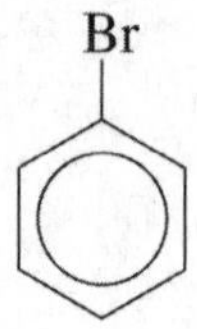

3.

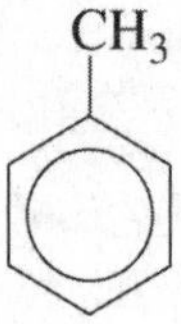

4.

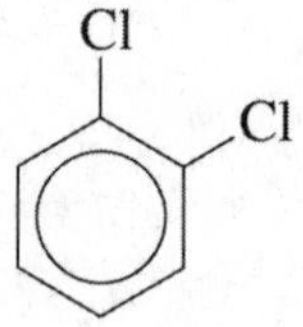

5.

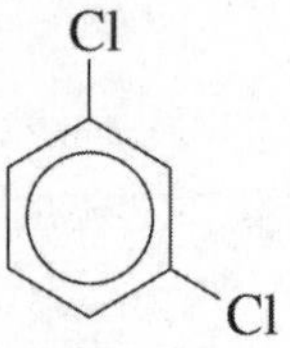

6.

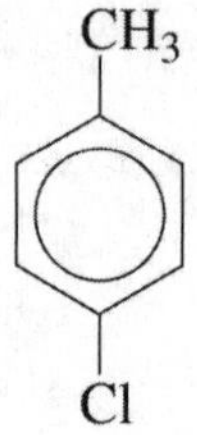

Answers

1. benzene
4. 1,2-dichlorobenzene

2. bromobenzene
5. 1,3-dichlorobenzene

3. methylbenzene; toluene
6. 4-chlorotoluene

Checklist for Chapter 11

You are ready to take the Practice Test for Chapter 11. Be sure that you have accomplished the following learning goals for this chapter. If you are not sure, review the section listed at the end of the goal. Then apply your new skills and understanding to the Practice Test.

After studying Chapter 11, I can successfully:

_____ Identify the structural features of alkenes and alkynes (11.1).

_____ Name alkenes and alkynes using IUPAC rules and draw their condensed structural formulas (11.1).

_____ Identify alkenes that exist as cis–trans isomers; draw their condensed structural formulas and names (11.2).

_____ Draw the condensed structural formulas for the products of the addition of hydrogen, halogens, and water to alkenes (11.3).

_____ Describe the process of forming polymers from alkene monomers (11.4).

_____ Give the names and draw the condensed structural formulas for compounds that contain a benzene ring (11.5).

Practice Test for Chapter 11

For questions 1 through 4, refer to the following compounds (A) and (B):

$$CH_2$$
$$H_2C=CH-CH_3 \qquad H_2C-CH_2$$

(A)   (B)

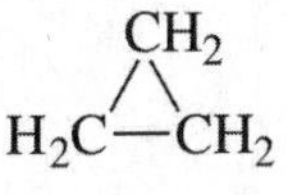

1. These compounds are
   A. aromatic   **B.** alkanes   **C.** isomers   **D.** alkenes   **E.** cycloalkanes

2. Compound (A) is a(n)
   A. alkane   **B.** alkene   **C.** cycloalkane   **D.** alkyne   **E.** aromatic

3. Compound (B) is named
   A. propane   **B.** propylene   **C.** cyclobutane   **D.** cyclopropane   **E.** cyclopropene

4. Compound (A) is named
   A. propane   **B.** propene   **C.** 2-propene   **D.** propyne   **E.** 1-butene

In questions 5 through 8, match each formula with its name:

   A. cyclopentene   **B.** methylpropene   **C.** cyclohexene
   D. ethene   **E.** 3-methylcyclopentene

5. $CH_2=CH_2$

6. 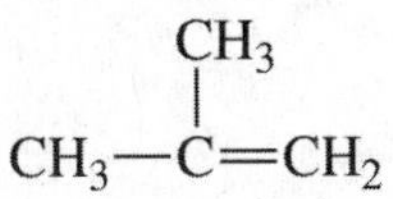
$$CH_3$$
$$CH_3-C=CH_2$$

7. 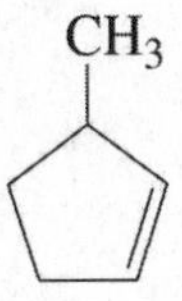
$$CH_3$$

8.

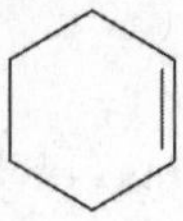

9. The cis isomer of 2-butene is

    A. CH_2=CH—CH_2—CH_3

    B. CH_3—CH=CH—CH_3

    C. 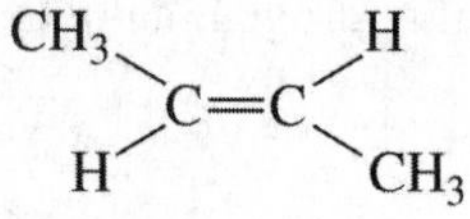
$$CH_3\!\!\diagdown \qquad \diagup H$$
$$C{=}C$$
$$H\!\!\diagup \qquad \diagdown CH_3$$

    D. 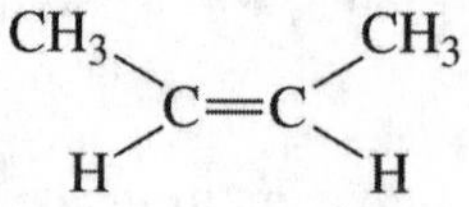
$$CH_3\!\!\diagdown \qquad \diagup CH_3$$
$$C{=}C$$
$$H\!\!\diagup \qquad \diagdown H$$

    E. 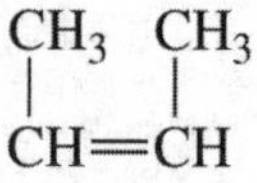
$$\overset{\displaystyle CH_3}{|} \quad \overset{\displaystyle CH_3}{|}$$
$$CH{=}CH$$

10. The name of this compound is

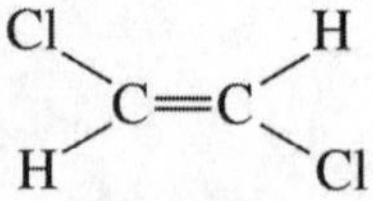

$$Cl\!\!\diagdown \qquad \diagup H$$
$$C{=}C$$
$$H\!\!\diagup \qquad \diagdown Cl$$

    A. dichloroethene
    B. *cis*-1,2-dichloroethene

    C. *trans*-1,2-dichloroethene
    D. *cis*-chloroethene

    E. *trans*-chloroethene

11. Hydrogenation of CH_3—CH=CH_2 gives

    A. $3CO_2 + 6H_2$
    B. CH_3—CH_2—CH_3
    C. CH_2=CH—CH_3

    D. no reaction
    E. CH_3—CH_2—CH_2—CH_3

12. Choose the product of the following reaction: CH_3—CH=CH_2 + HOH $\xrightarrow{\ H^+\ }$

    A. CH_3—CH_2—CH_2—OH
    B. no reaction
    C. CH_3—CH_2—CH_3

    D. 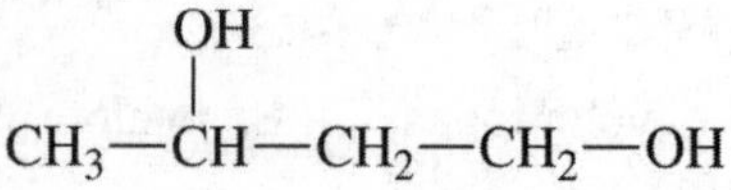 CH_3—$\overset{\overset{\textstyle OH}{|}}{C}H$—$CH_2$—$CH_2$—OH
    E. 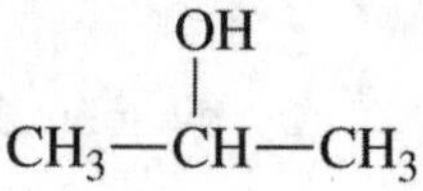 CH_3—$\overset{\overset{\textstyle OH}{|}}{C}H$—$CH_3$

13. Addition of bromine (Br_2) to ethene gives

    A. CH_3—CH_2—Br
    B. Br—CH_2—CH_2—Br
    C. CH_3—CH—Br_2

    D. CH_3—CH_3
    E. no reaction

14. Hydration of 2-butene gives

    A. CH_3—CH_2—CH_2—CH_3
    B. CH_3—CH_2—CH_2—CH_2—OH

    C. 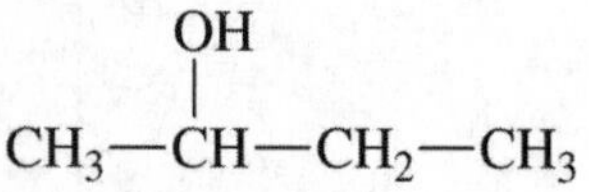 CH_3—$\overset{\overset{\textstyle OH}{|}}{C}H$—$CH_2$—$CH_3$
    D. ▢
    E. 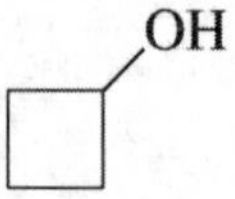 ▢⁻OH

15. What is the common name of methylbenzene?

    A. aniline
    B. phenol
    C. toluene

    D. xylene
    E. toluidine

16. What is the IUPAC name of CH_3—CH_2—C≡CH?

    A. methylacetylene
    B. propyne
    C. propylene

    D. 4-butyne
    E. 1-butyne

17. What is the product when cyclopentene reacts with Cl_2?
    A. chlorocyclopentene   **B.** 1,1-dichlorocyclopentane
    C. 1,2-dichlorocyclopentane   **D.** 1,3-dichlorocyclopentane
    E. no reaction

18. The reaction $CH_2{=}CH_2 + Cl_2 \rightarrow Cl{-}CH_2{-}CH_2{-}Cl$ is called:
    A. hydrogenation   **B.** halogenation   **C.** oxidation
    D. hydration   **E.** combustion

19. The following reaction is called:

$$CH_3{-}CH{=}CH_2 + H_2O \xrightarrow{H^+} CH_3{-}\underset{\underset{OH}{|}}{CH}{-}CH_3$$

    A. hydrogenation   **B.** halogenation   **C.** reduction
    D. hydration   **E.** combustion

For questions 20 through 23, match the compounds with one of the following families:

    A. alkane   **B.** alkene   **C.** alkyne   **D.** cycloalkene

20. $CH_3{-}CH{=}CH_2$

21.

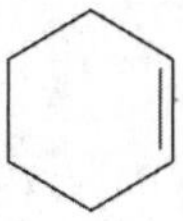

22. $CH_3{-}CH_2{-}\underset{\overset{|}{CH}}{\overset{CH_3}{|}}{-}CH_2{-}CH_3$

23. $CH_3{-}CH_2{-}C{\equiv}CH$

For questions 24 through 28, match the name of each of the following aromatic compounds with the correct structure.

A.   **B.**   **C.**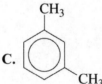

D. 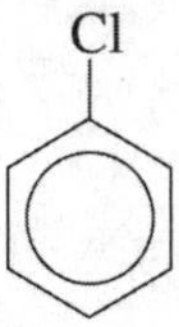   **E.**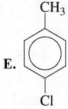

24. _____ chlorobenzene

25. _____ benzene

26. _____ toluene

27. _____ 4-chlorotoluene

28. _____ 1,3-dimethylbenzene

Answers to the Practice Test

1. C	**2.** B	**3.** D	**4.** B	**5.** D
6. B	**7.** E	**8.** C	**9.** D	**10.** C
11. B	**12.** E	**13.** B	**14.** C	**15.** C
16. E	**17.** C	**18.** B	**19.** D	**20.** B
21. D	**22.** A	**23.** C	**24.** D	**25.** A
26. B	**27.** E	**28.** C		

12

Organic Compounds with Oxygen and Sulfur

Study Goals

- Name and draw the condensed structural formulas for alcohols, phenols, and thiols.
- Name and draw the condensed structural formulas for ethers.
- Classify alcohols as primary, secondary, or tertiary.
- Describe the solubility in water of alcohols, phenols, and ethers.
- Write equations for combustion, dehydration, and oxidation of alcohols.
- Name and draw the condensed structural formulas for aldehydes and ketones.
- Describe the solubility of aldehydes and ketones in water.
- Write equations for the oxidation and reduction of aldehydes and ketones.
- Identify the chiral carbon atoms in organic molecules.

Think About It

1. What are the functional groups of alcohols, phenols, ethers, and thiols?

2. Phenol is sometimes used in mouthwashes. Why does it form a solution with water?

3. What reaction of ethanol takes place when you make a flambé dessert?

4. What are the functional groups of an aldehyde and ketone?

5. How is an alcohol changed to an aldehyde or ketone?

Key Terms

Match the following terms with the statements shown below.

- **a.** primary alcohol
- **b.** thiol
- **c.** ether
- **d.** phenol
- **e.** tertiary alcohol
- **f.** chiral carbon
- **g.** aldehyde
- **h.** ketone

1. _____ an organic compound with one alkyl group bonded to the carbon with the —OH group

2. _____ an organic compound that contains an —SH group

3. _____ an organic compound that contains an oxygen atom —O— attached to two alkyl groups

4. _____ an organic compound with three alkyl groups bonded to the carbon with the —OH group

5. _____ an organic compound that contains a benzene ring bonded to a hydroxyl group

6. _____ an organic compound with a carbonyl group attached to two alkyl groups

7. _____ a carbon that is bonded to four different groups

8. _____ an organic compound that contains a carbonyl group and a hydrogen atom at the end of the carbon chain

Answers   **1.** a   **2.** b   **3.** c   **4.** e   **5.** d   **6.** h   **7.** f   **8.** g

12.1 Alcohols, Phenols, Thiols, and Ethers

- An alcohol contains the hydroxyl group —OH attached to a carbon chain.
- In the IUPAC system alcohols are named by replacing the *ane* of the alkane name with *ol*. The location of the —OH group is given by numbering the carbon chain. Simple alcohols are generally named by their common names with the alkyl name preceding the term *alcohol*. For example, CH_3—OH is methyl alcohol, and CH_3—CH_2—OH is ethyl alcohol.

    CH_3—OH CH_3—CH_2—OH CH_3—CH_2—CH_2—OH
    Methanol   Ethanol   1-Propanol
    (methyl alcohol)   (ethyl alcohol)   (propyl alcohol)

- When a hydroxyl group is attached to a benzene ring, the compound is a phenol.
- Thiols are similar to alcohols, except they have an —SH functional group in place of the —OH group.
- In ethers, an oxygen atom is connected by two single bonds to alkyl or aromatic groups.
- In the common names of ethers, the alkyl groups are listed alphabetically followed by the name *ether*.

(MC)

Tutorial: Naming Alcohols, Phenols, and Thiols
Tutorial: Drawing Alcohols, Phenols, and Thiols

Guide to Naming Alcohols	
Step 1	Name the longest carbon chain with the —OH group. Name an aromatic alcohol as a *phenol*.
Step 2	Number the chain starting at the end closer to the —OH.
Step 3	Give the location and name of any substituent relative to the —OH group.

♦ **Learning Exercise 12.1A**

Give the correct IUPAC and common name (if any) for each of the following compounds:

1. $CH_3—CH_2—OH$

2. $CH_3—CH_2—CH_2—OH$

3.
$$\overset{\displaystyle OH}{\underset{\displaystyle |}{CH_3—CH—CH_2—CH_2—CH_3}}$$

4.

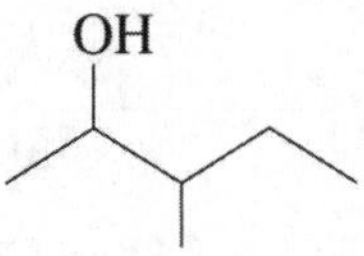

5.

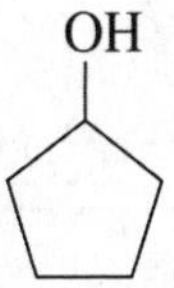

6.

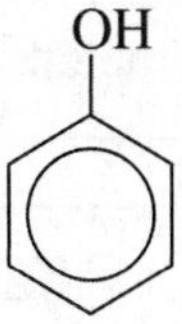

Answers

1. ethanol (ethyl alcohol)
3. 2-pentanol
5. cyclopentanol

2. 1-propanol (propyl alcohol)
4. 3-methyl-2-pentanol
6. phenol

♦ **Learning Exercise 12.1B**

Draw the correct condensed structural formula for **1.-3.**, for **4.-6.**, draw the skeletal formula.

1. 2-butanol

2. 2-chloro-1-propanol

3. 2,4-dimethyl-1-pentanol

4. 2-bromoethanol

5. cyclohexanol

6. 2-chlorophenol

173

Answers

$$\underset{\overset{|}{OH}}{CH_3-CH-CH_2-CH_3} \quad \textbf{2.} \quad \underset{\overset{|}{Cl}}{CH_3-CH-CH_2-OH} \quad \textbf{3.} \quad \underset{\overset{|}{CH_3}}{CH_3-CH-CH_2-\underset{\overset{|}{CH_3}}{CH}-CH_2-OH}$$

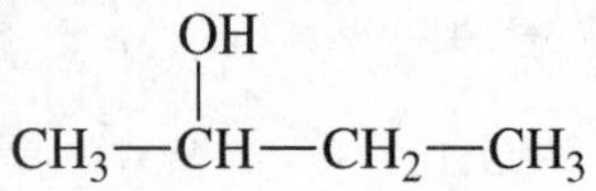

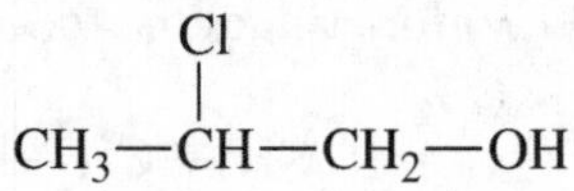

1.

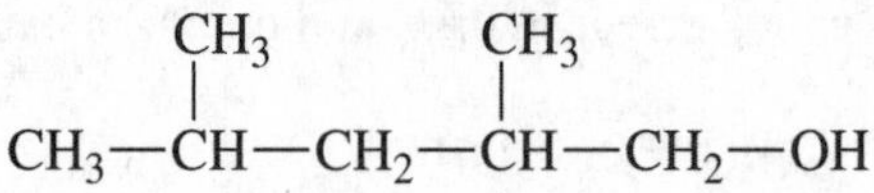

4. 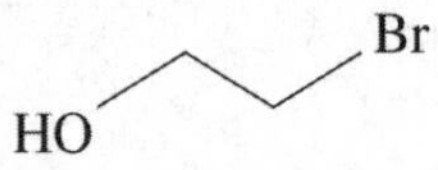 HO—CH₂—CH₂—Br

5. 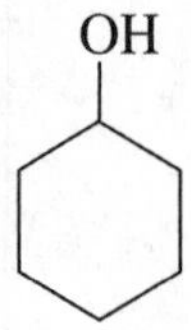 (cyclohexanol, OH)

6. 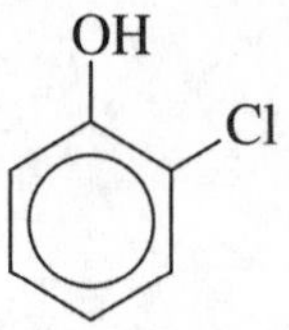 (phenol with Cl)

Study Note

Example: Write the common name for $CH_3-CH_2-O-CH_3$.

Solution: The common name lists the alkyl groups alphabetically before the name *ether*.

Ethyl group   Methyl group

$$CH_3-CH_2-O-CH_3$$

Common: ethyl methyl ether

(MC)

Tutorial: Naming Ethers
Tutorial: Drawing Ethers

♦ **Learning Exercise 12.1C**

Write the common name for each of the following ethers:

1. CH_3-O-CH_3 **2.** $CH_3-CH_2-O-CH_2-CH_3$

3. $CH_3-CH_2-CH_2-CH_2-O-CH_3$ **4.** $CH_3-O-CH_2-CH_3$

Answers   **1.** (di)methyl ether   **2.** (di)ethyl ether
                3. butyl methyl ether   **4.** ethyl methyl ether

♦ **Learning Exercise 12.1D**

Draw the condensed structural formula for each of the following ethers:

1. ethyl propyl ether   **2.** dibutyl ether

Answers   **1.** $CH_3-CH_2-O-CH_2-CH_2-CH_3$
                2. $CH_3-CH_2-CH_2-CH_2-O-CH_2-CH_2-CH_2-CH_3$

♦ **Learning Exercise 12.1E**

Identify each of the following pairs of compounds as isomers or the same compound:

1. CH_3-O-CH_3 and CH_3-CH_2-OH _____

2. $CH_3-O-CH_2-CH_3$ and $CH_3-\overset{\overset{\displaystyle OH}{|}}{CH}-CH_3$ _____

3. $CH_3-\overset{\overset{\displaystyle CH_3}{|}}{CH}-OH$ and $CH_3-\overset{\overset{\displaystyle OH}{|}}{CH}-CH_3$ _____

Answers   1. isomers of C_2H_6O 2. isomers of C_3H_8O
             3. same compound (2-propanol)

12.2 Properties of Alcohols and Ethers

- Alcohols are classified according to the number of alkyl groups attached to the carbon bonded to the —OH group.
- In a primary alcohol, there is one alkyl or aromatic group attached to the carbon atom bonded to the —OH. In a secondary alcohol, there are two alkyl or aromatic groups, and in a tertiary alcohol there are three alkyl or aromatic groups attached to the carbon atom with the —OH functional group.
- Alcohols with one to four carbons are soluble in water because the —OH group forms hydrogen bonds with water molecules.
- Phenol is soluble in water and acts as a weak acid.
- Ethers are widely used as solvents but can be dangerous to use because their vapors are highly flammable.

Study Note

Example: Identify the following as primary, secondary, or tertiary alcohols.
Solution: Determine the number of alkyl groups attached to the hydroxyl carbon atoms.

$$CH_3-CH_2-OH \qquad CH_3-\overset{\overset{\displaystyle CH_3}{|}}{CH}-OH \qquad CH_3-\overset{\overset{\displaystyle CH_3}{|}}{\underset{\underset{\displaystyle CH_3}{|}}{C}}-OH$$

Primary (1°)   Secondary (2°)   Tertiary (3°)

♦ **Learning Exercise 12.2A**

Classify each of the following alcohols as primary (1°), secondary (2°), or tertiary (3°):

1. _____ CH_3-CH_2-OH

2. _____ (structure with OH on central carbon)

3. _____ $CH_3-\overset{\overset{\displaystyle OH}{|}}{\underset{\underset{\displaystyle CH_3}{|}}{C}}-CH_2-CH_3$

4. _____ $CH_3-\overset{\overset{\displaystyle OH}{|}}{\underset{\underset{\displaystyle CH_3}{|}}{C}}-CH_2-CH_2-CH_3$

5. _____ CH₃ structure with CH₃ group

$$CH_3-\overset{\overset{\displaystyle CH_3}{|}}{\underset{\underset{\displaystyle CH_3}{|}}{C}}-CH_2-OH$$

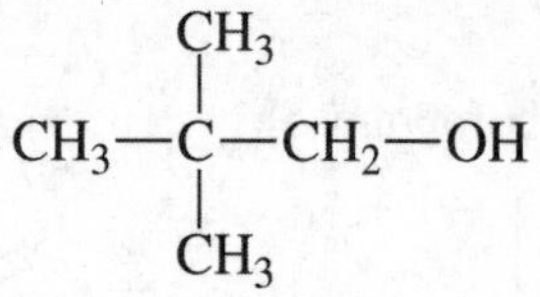

6. _____

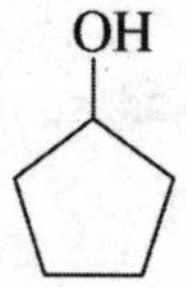

Answers
1. primary (1°) 2. secondary (2°) 3. tertiary (3°)
4. tertiary (3°) 5. primary (1°) 6. secondary (2°)

♦ **Learning Exercise 12.2B**

Select the compound in each pair that is the more soluble in water.

1. CH_3-CH_3 or CH_3-CH_2-OH

2. $CH_3-CH_2-CH_2-OH$ or $CH_3-CH_2-CH_2-CH_2-CH_2-OH$

3. $CH_3-CH_2-CH_2-CH_3$ or $CH_3-CH_2-CH_2-CH_2-OH$

4. benzene or phenol

Answers
1. CH_3-CH_2-OH 2. $CH_3-CH_2-CH_2-OH$
3. $CH_3-CH_2-CH_2-CH_2-OH$ 4. phenol

12.3 Reactions of Alcohols and Thiols

- At high temperatures, an alcohol dehydrates in the presence of an acid to yield an alkene and water.

$$CH_3-CH_2-OH \xrightarrow[\text{Heat}]{H^+} H_2C{=}CH_2 + H_2O$$

- Using an oxidizing agent [O], primary alcohols oxidize to aldehydes, which usually oxidize further to carboxylic acids. Secondary alcohols are oxidized to ketones, but tertiary alcohols do not oxidize.

$$CH_3-CH_2-OH \xrightarrow{[O]} CH_3-\overset{\overset{\displaystyle O}{\|}}{C}-H + H_2O$$

1° Alcohol   Aldehyde

$$CH_3-\overset{\overset{\displaystyle OH}{|}}{C}H-CH_3 \xrightarrow{[O]} CH_3-\overset{\overset{\displaystyle O}{\|}}{C}-CH_3 + H_2O$$

2° Alcohol   Ketone

MC

Case Study: Alcohol Toxicity
Tutorial: Dehydration and Oxidation of Alcohols

♦ **Learning Exercise 12.3A**

Draw the condensed structural formulas or skeletal formulas of the products expected from dehydration of each of the following reactants:

1. $CH_3-CH_2-CH_2-CH_2-OH$ $\xrightarrow{H^+, \text{ heat}}$

2. 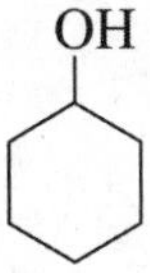 $\xrightarrow{H^+, \text{ heat}}$

3. 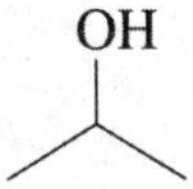 $\xrightarrow{H^+, \text{ heat}}$

4. $CH_3-CH_2-\overset{\overset{\displaystyle OH}{|}}{CH}-CH_2-CH_3$ $\xrightarrow{H^+, \text{ heat}}$

Answers

1. $CH_3-CH_2-CH=CH_2$ 2.

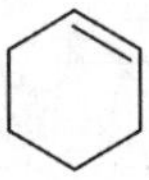

3. 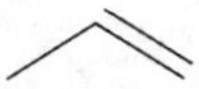   4. $CH_3-CH_2-CH=CH-CH_3$

♦ **Learning Exercise 12.3B**

Draw the condensed structural formulas or skeletal formulas of the aldehyde or ketone expected in the oxidation of each of the following reactants:

1. $CH_3-CH_2-CH_2-CH_2-OH$ $\xrightarrow{[O]}$

2. 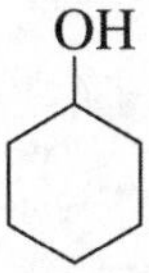 $\xrightarrow{[O]}$

3. 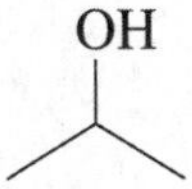 $\xrightarrow{[O]}$

4. $CH_3-CH_2-\overset{\overset{\displaystyle OH}{|}}{CH}-CH_2-CH_3$ $\xrightarrow{[O]}$

Answers

1. $CH_3-CH_2-CH_2-\overset{\overset{\displaystyle O}{||}}{C}-H$ 2.

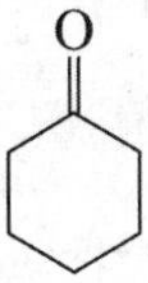

3. 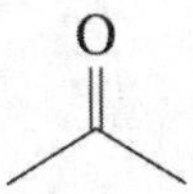   4. $CH_3-CH_2-\overset{\overset{\displaystyle O}{||}}{C}-CH_2-CH_3$

177

12.4 Aldehydes and Ketones

- In an aldehyde, the carbonyl group appears at the end of a carbon chain attached to at least one hydrogen atom.
- In a ketone, the carbonyl group occurs between carbon groups and has no hydrogens attached to it.
- In the IUPAC system, aldehydes and ketones are named by replacing the *e* in the longest chain containing the carbonyl group with *al* for aldehydes, and *one* for ketones. The location of the carbonyl group in a ketone is given if there are more than four carbon atoms in the chain.

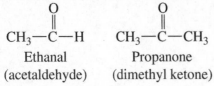

Ethanal   Propanone
(acetaldehyde)   (dimethyl ketone)

Tutorial: Aldehydes or Ketone?
Tutorial: Naming Aldehydes and Ketones

Guide to Naming Aldehydes	
Step 1	Name the longest carbon chain containing the carbonyl group by replacing the *e* in the alkane name with *al*.
Step 2	Name and number substituents by counting the carbonyl group as carbon 1.

◆ Learning Exercise 12.4A

Classify each of the following compounds:

    a. alcohol   **b.** aldehyde   **c.** ketone   **d.** ether   **e.** thiol

1. _____ $CH_3-CH_2-CH_2-\overset{\overset{\displaystyle O}{\|}}{C}-H$

2. _____ (skeletal structure with OH)

3. _____ $CH_3-CH_2-\overset{\overset{\displaystyle O}{\|}}{C}-CH_2-CH_3$

4. _____ $CH_3-CH_2-O-CH_3$

5. _____ $CH_3-\overset{\overset{\displaystyle O}{\|}}{C}-CH_2-CH_3$

6. _____ $CH_3-\overset{\overset{\displaystyle O}{\|}}{C}-H$

7. _____ $CH_3-CH_2-\overset{\overset{\displaystyle SH}{|}}{CH}-CH_3$

8. _____ $CH_3-CH_2-\overset{\overset{\displaystyle OH}{|}}{CH}-CH_3$

Answers   **1.** b   **2.** a   **3.** c   **4.** d
                        5. c   **6.** b   **7.** e   **8.** a

♦ **Learning Exercise 12.4B**

Write the correct IUPAC name and common name, if any, for the following aldehydes:

1. $CH_3-\overset{\overset{O}{\|}}{C}-H$ 2. $CH_3-CH_2-CH_2-CH_2-\overset{\overset{O}{\|}}{C}-H$

3. $CH_3-CH_2-\overset{\overset{\displaystyle CH_3}{|}}{CH}-CH_2-CH_2-\overset{\overset{O}{\|}}{C}-H$ 4.

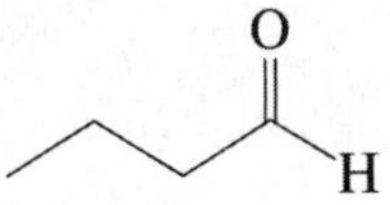

5. $H-\overset{\overset{O}{\|}}{C}-H$

Answers   1. ethanal; acetaldehyde   2. pentanal   3. 4-methylhexanal
              4. butanal; butyraldehyde   5. methanal; formaldehyde

Guide to Naming Ketones	
Step 1	Name the longest carbon chain that contains the carbonyl group by replacing the *e* in the alkane name with *one*.
Step 2	Number the carbon chain starting from the end nearer the carbonyl group and indicate its position.
Step 3	Name and number any substituents on the carbon chain.

♦ **Learning Exercise 12.4C**

Write the IUPAC name and common name, if any, for the following ketones:

1. $CH_3-\overset{\overset{O}{\|}}{C}-CH_3$ 2.

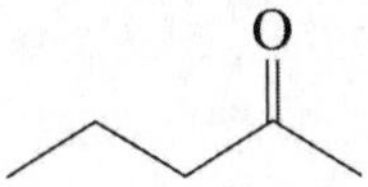

3. $CH_3-CH_2-\overset{\overset{O}{\|}}{C}-CH_2-CH_3$ 4.

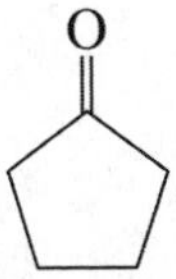

Answers   1. propanone; dimethyl ketone, acetone   2. 2-pentanone; methyl propyl ketone
              3. 3-pentanone; diethyl ketone   4. cyclopentanone

♦ **Learning Exercise 12.4D**

Draw the condensed structural formulas for **1** to **4** and the skeletal formula for **5** and **6**.

1. ethanal

2. 2-methylbutanal

3. 2-chloropropanal

4. ethylmethylketone

5. 3-hexanone

6. benzaldehyde

Answers

1. CH$_3$—C(=O)—H

2. CH$_3$—CH$_2$—CH(CH$_3$)—C(=O)—H

3. CH$_3$—CH(Cl)—C(=O)—H

4. CH$_3$—CH$_2$—C(=O)—CH$_3$

5.

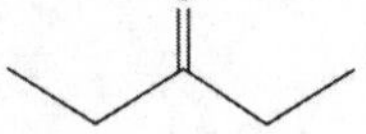

6.

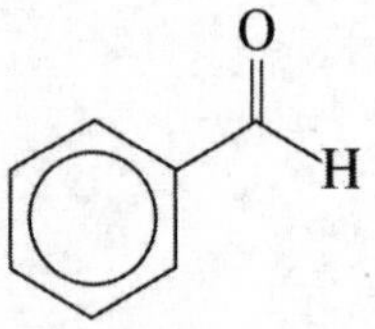

12.5 Properties of Aldehydes and Ketones

- The polarity of the carbonyl group makes aldehydes and ketones of one to four carbon atoms soluble in water.
- Using an oxidizing agent, aldehydes oxidize to carboxylic acids. Ketones do not oxidize further.

$$CH_3-\overset{\overset{\displaystyle O}{\|}}{C}-H \xrightarrow{[O]} CH_3-\overset{\overset{\displaystyle O}{\|}}{C}-OH$$

   Aldehyde   Carboxylic acid

$$CH_3-\overset{\overset{\displaystyle O}{\|}}{C}-CH_3 \xrightarrow{[O]} \text{No reaction}$$

   Ketone

- Aldehydes and ketones are reduced when hydrogen is added in the presence of a metal catalyst to produce primary or secondary alcohols.

(MC)

Tutorial: Oxidation–Reduction Reactions of Aldehydes and Ketones

---

◆ **Learning Exercise 12.5A**

Indicate whether each of the following compounds is soluble (S) or not soluble (NS) in water:

1. _____ 3-hexanone   2. _____ propanal   3. _____ acetaldehyde

4. _____ butanal   5. _____ cyclohexanone

Answers   1. NS   2. S   3. S   4. S   5. NS

◆ **Learning Exercise 12.5B**

Indicate the compound in each of the following pairs that will oxidize:

1. propanal or propanone   _____

2. butane or butanal   _____

3. ethane or acetaldehyde   _____

Answers   1. propanal   2. butanal   3. acetaldehyde

◆ **Learning Exercise 12.5C**

Draw the condensed structural formulas for the reduction products from each of the following:

1. $CH_3-\overset{\overset{\displaystyle O}{\|}}{C}-CH_3 + H_2 \xrightarrow{Pt}$

2. $CH_3-CH_2-\overset{\overset{\displaystyle O}{\|}}{C}-H + H_2 \xrightarrow{Pt}$

3. $CH_3 — \overset{\overset{\displaystyle O}{\|}}{C} — H + H_2 \xrightarrow{Pt}$

4. 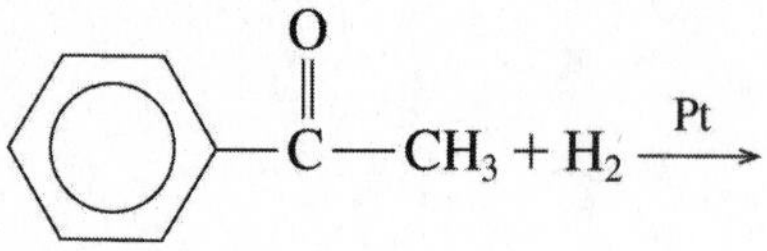 ⬡ $— \overset{\overset{\displaystyle O}{\|}}{C} — CH_3 + H_2 \xrightarrow{Pt}$

5. 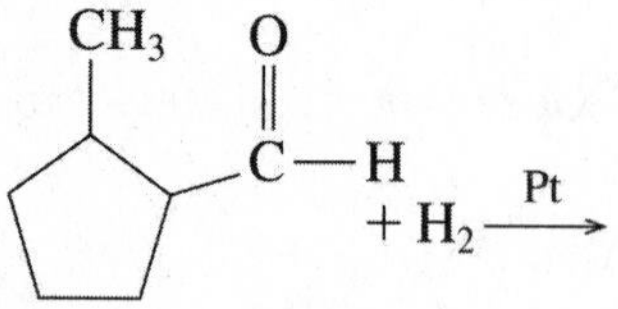 [cyclopentane with CH_3 and $\overset{\overset{\displaystyle O}{\|}}{C} — H$] $+ H_2 \xrightarrow{Pt}$

Answers

1. $CH_3 — \overset{\overset{\displaystyle OH}{|}}{CH} — CH_3$

2. $CH_3 — CH_2 — CH_2 — OH$

3. $CH_3 — CH_2 — OH$

4. 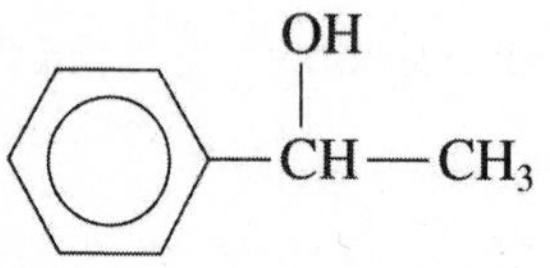 ⬡ $— \overset{\overset{\displaystyle OH}{|}}{CH} — CH_3$

5. 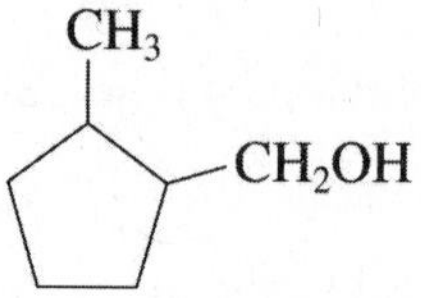 [cyclopentane with CH_3 and CH_2OH]

12.6 Chiral Molecules

- Chiral molecules have mirror images that cannot be superimposed.
- In a chiral molecule, there is one or more carbon atoms attached to four different atoms or groups.
- The mirror images of a chiral molecule represent two different molecules called enantiomers.
- In a Fischer projection (straight chain), the prefixes D- and L- are used to distinguish between the mirror images. In D-glyceraldehyde, the —OH is on the right of the chiral carbon; it is on the left in L-glyceraldehyde.

Tutorial: Chiral Carbon Atoms

♦ **Learning Exercise 12.6A**

Indicate whether the following objects would be chiral or not:

1. _____ a piece of plain paper

2. _____ a weight-lifting glove

3. _____ a baseball cap

4. _____ a volleyball net

5. _____ your left foot

Answers   1. not chiral   2. chiral   3. not chiral   4. not chiral   5. chiral

♦ **Learning Exercise 12.6B**

State whether each of the following molecules is chiral or not chiral:

$$
\begin{array}{ccc}
& Cl & \\
& | & \\
H\!-\!\!C\!-\!Cl & \\
& | & \\
\mathbf{1.} & CH_3 &
\end{array}
\qquad
\begin{array}{ccc}
& Cl & \\
& | & \\
H\!-\!\!C\!-\!OH & \\
& | & \\
\mathbf{2.} & CH_3 &
\end{array}
\qquad
\begin{array}{ccc}
& CHO & \\
& | & \\
H\!-\!\!C\!-\!OH & \\
& | & \\
\mathbf{3.} & CH_3 &
\end{array}
$$

_____   _____   _____

Answers   **1.** not chiral   **2.** chiral   **3.** chiral

♦ **Learning Exercise 12.6C**

Identify the following as a feature that is characteristic of a chiral compound or not:

 1. the central atom is attached to two identical groups   _____

 2. contains a carbon attached to four different groups   _____

 3. has identical mirror images   _____

Answers   **1.** not chiral   **2.** chiral   **3.** not chiral

♦ **Learning Exercise 12.6D**

Indicate whether each pair of Fischer projections represents enantiomers (E) or identical structures (I).

1.
$$
\begin{array}{c}
COOH \\
HO\!-\!\!\!-\!\!H \\
CH_2OH
\end{array}
\quad and \quad
\begin{array}{c}
COOH \\
H\!-\!\!\!-\!\!OH \\
CH_2OH
\end{array}
$$

2.
$$
\begin{array}{c}
CH_2OH \\
Cl\!-\!\!\!-\!\!H \\
CH_3
\end{array}
\quad and \quad
\begin{array}{c}
CH_2OH \\
Cl\!-\!\!\!-\!\!H \\
CH_3
\end{array}
$$

3.
$$
\begin{array}{c}
CHO \\
Cl\!-\!\!\!-\!\!Br \\
CH_3
\end{array}
\quad and \quad
\begin{array}{c}
CHO \\
Br\!-\!\!\!-\!\!Cl \\
CH_3
\end{array}
$$

4.
$$
\begin{array}{c}
COOH \\
H\!-\!\!\!-\!\!OH \\
COOH
\end{array}
\quad and \quad
\begin{array}{c}
COOH \\
HO\!-\!\!\!-\!\!H \\
COOH
\end{array}
$$

Answers   **1.** E   **2.** I   **3.** E   **4.** I

Checklist for Chapter 12

You are ready to take the Practice Test for Chapter 12. Be sure that you have accomplished the following learning goals for this chapter. If you are not sure, review the section listed at the end of the goal. Then apply your new skills and understanding to the Practice Test.

After studying Chapter 12, I can successfully:

_____ Give the IUPAC or common name of an alcohol (12.1).

_____ Classify an alcohol as primary, secondary, or tertiary (12.2).

_____ Describe the solubility of alcohols, phenols, and ethers in water (12.2).

_____ Draw the condensed structural formulas for reactants and products of the oxidation of alcohols (12.3).

_____ Identify condensed structural formulas as aldehydes and ketones (12.4).

_____ Give the IUPAC and common names of an aldehyde or ketone; draw the condensed structural formula from the name (12.4).

_____ Describe the solubility of aldehydes and ketones in water (12.5).

_____ Draw the condensed structural formulas for the oxidation and reduction products of aldehydes and ketones (12.5).

_____ Identify a molecule as a chiral or not (12.6).

Practice Test for Chapter 12

For questions 1 through 5, match the names with the correct structure.

A. 1-propanol   **B.** cyclobutanol   **C.** 2-propanol   **D.** ethyl methyl ether   **E.** diethyl ether

1.
$$CH_3 - \overset{\overset{\displaystyle OH}{|}}{CH} - CH_3$$

2. $CH_3 - CH_2 - CH_2 - OH$

3. $CH_3 - O - CH_2 - CH_3$

4. ☐—OH

5. $CH_3 - CH_2 - O - CH_2 - CH_3$

6. The compound $CH_3 - \overset{\overset{\displaystyle O}{||}}{C} - CH_3$ is formed by the oxidation of
  A. 2-propanol   **B.** propane   **C.** 1-propanol
  D. dimethyl ether   **E.** methyl ethyl ketone

7. Why are short-chain alcohols water soluble?
  A. They are nonpolar.   **B.** They can hydrogen bond.   **C.** They are organic.
  D. They are bases.   **E.** They are acids.

8. Phenol is
  A. the alcohol of benzene   **B.** the aldehyde of benzene   **C.** the phenyl group of benzene
  D. the ketone of benzene   **E.** cyclohexanol

9.
$$CH_3 - \overset{\overset{\displaystyle OH}{|}}{CH} - CH_3 \xrightarrow{[O]}$$
  A. an alkane   **B.** an aldehyde   **C.** a ketone   **D.** an ether   **E.** a phenol

10. The dehydration of cyclohexanol gives
  A. cyclohexane   **B.** cyclohexene   **C.** cyclohexyne   **D.** benzene   **E.** phenol

11. The condensed structural formula of the thiol of ethane is

    A. CH_3—SH   **B.** CH_3—CH_2—OH   **C.** CH_3—CH_2—SH

    D. CH_3—CH_2—S—CH_3 **E.** CH_3—S—OH

For questions 12 through 16, classify each alcohol as

    A. primary (1°)   **B.** secondary (2°)   **C.** tertiary (3°)

12. CH_3—CH_2—CH_2—OH

13.

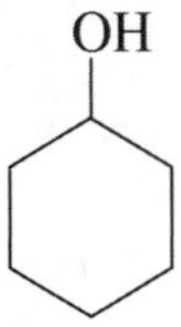

14. 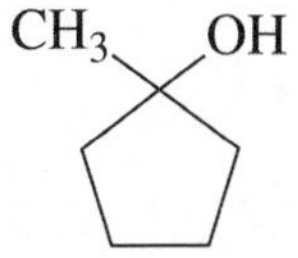
$$\text{a cyclopentane ring with } CH_3 \text{ and } OH \text{ on the same carbon}$$

15.
$$\begin{array}{c} \quad\quad OH \\ \quad\quad | \\ CH_3-\overset{\displaystyle |}{\underset{\displaystyle |}{C}}-CH_2-CH_2-CH_3 \\ \quad\quad CH_3 \end{array}$$

16.
$$\begin{array}{c} \quad\quad OH \\ \quad\quad | \\ CH_3-CH-CH_2-CH_2-CH_2-CH_3 \end{array}$$

Complete questions 17 through 20 by indicating the products (A–E) formed in each of the following reactions:

    A. primary alcohol   **B.** secondary alcohol   **C.** aldehyde   **D.** ketone   **E.** carboxylic acid

17. _____ oxidation of a primary alcohol

18. _____ oxidation of a secondary alcohol

19. _____ oxidation of an aldehyde

20. _____ reduction of an aldehyde

For questions 21 through 25, match the names with the correct structures.

    A. dimethyl ether   **B.** acetaldehyde   **C.** methanal   **D.** dimethyl ketone   **E.** propanal

21. _____
$$\begin{array}{c} \quad O \\ \quad || \\ H-C-H \end{array}$$

22. _____
$$\begin{array}{c} \quad\quad O \\ \quad\quad || \\ CH_3-C-H \end{array}$$

23. _____ CH_3—O—CH_3

24. _____ $CH_3—CH_2—\overset{\overset{\displaystyle O}{\|}}{C}—H$

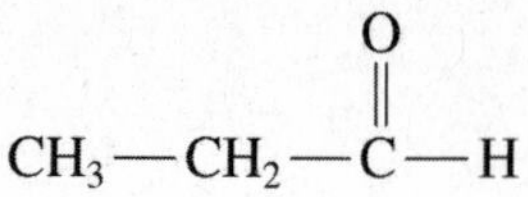

25. _____ $CH_3—\overset{\overset{\displaystyle O}{\|}}{C}—CH_3$

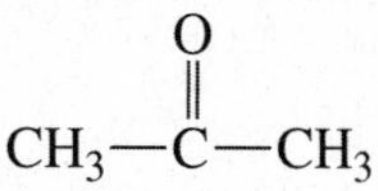

26. An isomer of $CH_3—CH_2—CH_2—OH$ is

    A. $HO—CH_2—CH_2—CH_3$

    B. 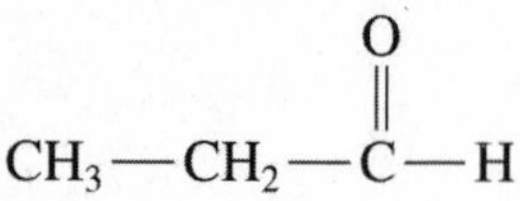 $CH_3—CH_2—\overset{\overset{\displaystyle O}{\|}}{C}—H$

    C. $CH_3—CH_2—O—CH_3$

    D. 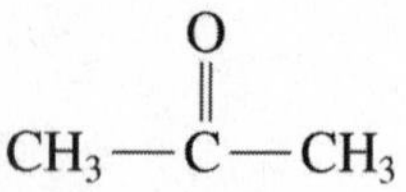 $CH_3—\overset{\overset{\displaystyle O}{\|}}{C}—CH_3$

    E. $CH_3—CH_2—CH_3$

Identify each of the following condensed structural formulas as

    A. alcohol   **B.** ether

27. 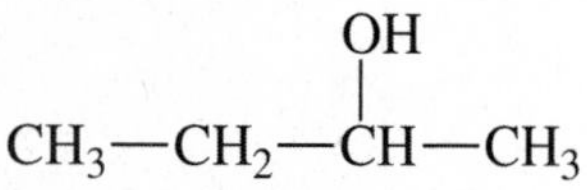 $CH_3—CH_2—\overset{\overset{\displaystyle OH}{|}}{CH}—CH_3$

28. 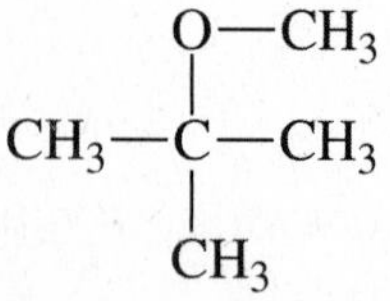 $CH_3—\overset{\overset{\displaystyle O—CH_3}{|}}{\underset{\underset{\displaystyle CH_3}{|}}{C}}—CH_3$

29. The condensed structural formula for 4-bromo-3-methylcyclohexanone is

  A. 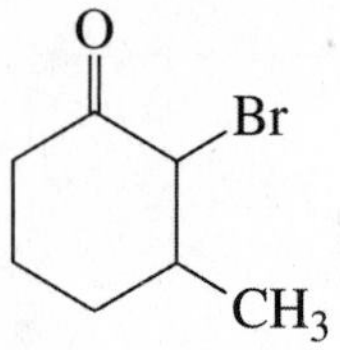   **B.** 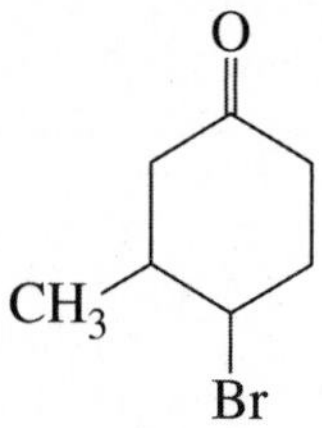   **C.**

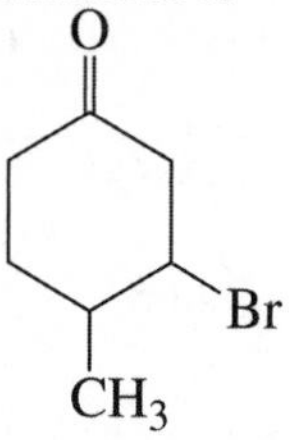

  D. 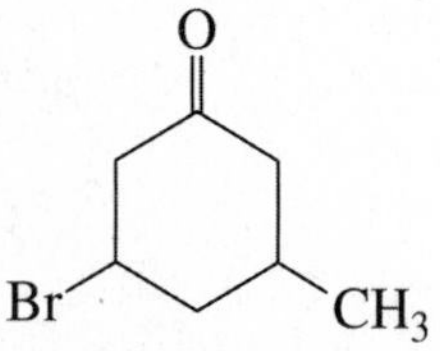   **E.**

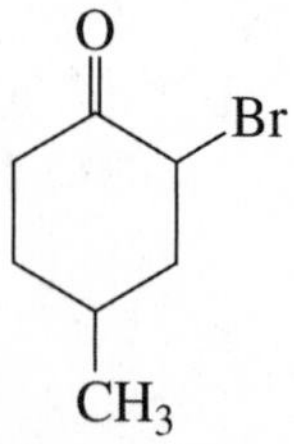

For questions 30 through 34, identify as enantiomers (E), or identical (I).

30. 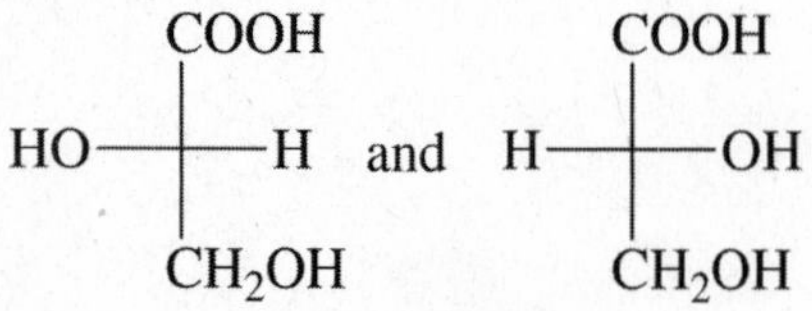 $HO—\overset{\overset{\displaystyle COOH}{|}}{\underset{\underset{\displaystyle CH_2OH}{|}}{}}—H$ and $H—\overset{\overset{\displaystyle COOH}{|}}{\underset{\underset{\displaystyle CH_2OH}{|}}{}}—OH$

31. 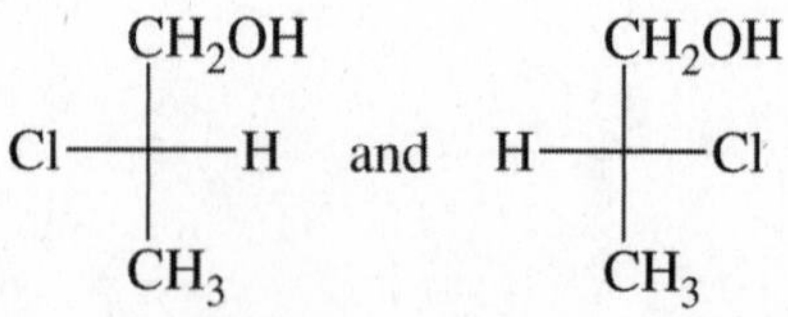 $Cl—\overset{\overset{\displaystyle CH_2OH}{|}}{\underset{\underset{\displaystyle CH_3}{|}}{}}—H$ and $H—\overset{\overset{\displaystyle CH_2OH}{|}}{\underset{\underset{\displaystyle CH_3}{|}}{}}—Cl$

32.

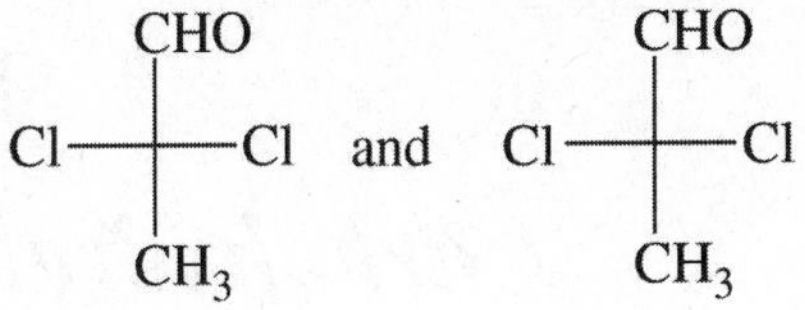

    CHO   CHO
Cl——Cl and Cl——Cl
    CH₃   CH₃

33.

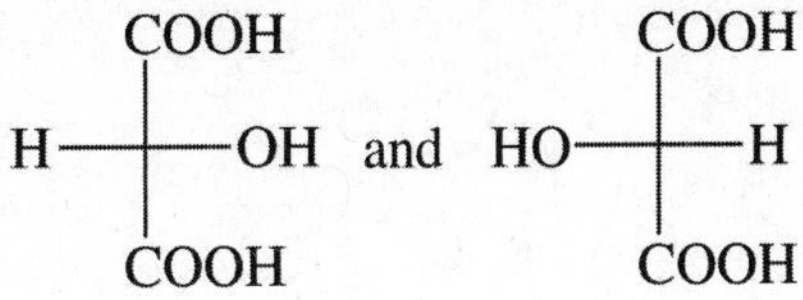

    COOH   COOH
H——OH and HO——H
    COOH   COOH

34.

    CHO   CHO
Cl——Br and Cl——Br
    CH₃   CH₃

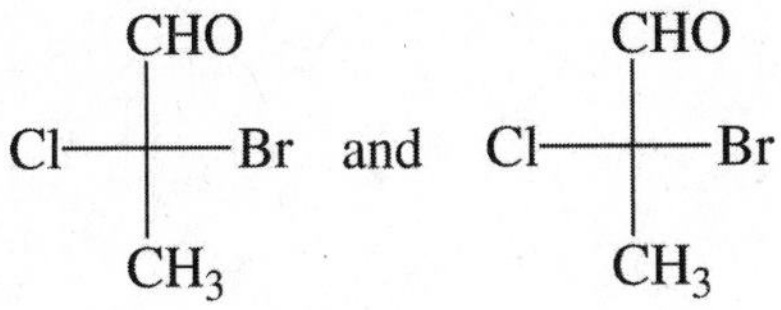

Answers to Practice Test

1. C	2. A	3. D	4. B	5. E
6. A	7. B	8. A	9. C	10. B
11. C	12. A	13. B	14. C	15. C
16. B	17. C, E	18. D	19. E	20. A
21. C	22. B	23. A	24. E	25. D
26. C	27. A	28. B	29. B	30. E
31. E	32. I	33. I	34. I	

Study Goals

- Identify the common carbohydrates in the diet.
- Distinguish between monosaccharides, disaccharides, and polysaccharides.
- Identify the chiral carbons in a carbohydrate.
- Label the Fischer projection for a monosaccharide as the D or L isomer.
- Draw Haworth structures for monosaccharides.
- Describe the structural units and bonds in disaccharides and polysaccharides.

Think About It

1. What are some foods that contain carbohydrates?

2. What elements are found in carbohydrates?

3. What carbohydrates are present in table sugar, milk, and wood?

4. What is meant by a "high-fiber" diet?

Key Terms

Match the following key terms with the descriptions:

    a. carbohydrate   **b.** glucose   **c.** disaccharide
    d. Haworth structure   **e.** cellulose

1. _____ a simple or complex sugar composed of a carbon chain with an aldehyde or ketone group and several hydroxyl groups

2. _____ a cyclic structure that represents the closed chain form of a monosaccharide

3. _____ an unbranched polysaccharide that cannot be digested by humans

4. _____ an aldohexose that is the most prevalent monosaccharide in the diet

5. _____ a carbohydrate that contains two monosaccharides linked by a glycosidic bond

Answers **1.** a **2.** d **3.** e **4.** b **5.** c

13.1 Carbohydrates

- Carbohydrates are classified as monosaccharides (simple sugars), disaccharides (two monosaccharide units), and polysaccharides (many monosaccharide units).
- In a chiral molecule, there is one or more carbon atoms attached to four different atoms or groups.
- Monosaccharides are polyhydroxy aldehydes (aldoses) or ketones (ketoses).
- Monosaccharides are classified by the number of carbon atoms as *trioses, tetroses, pentoses,* or *hexoses.*

(MC)

Self Study Activity: Carbohydrates
Tutorial: Types of Carbohydrates
Self Study Activity: Forms of Carbohydrates
Tutorial: Carbonyls in Carbohydrates
Tutorial: Carbohydrates

♦ Learning Exercise 13.1A

Complete and balance the equations for the photosynthesis of

1. Glucose: _____ + _____ $\rightarrow C_6H_{12}O_6$ + _____

2. Ribose: _____ + _____ $\rightarrow C_5H_{10}O_5$ + _____

Answers   **1.** $6CO_2 + 6H_2O \rightarrow C_6H_{12}O_6 + 6O_2$
          2. $5CO_2 + 5H_2O \rightarrow C_5H_{10}O_5 + 5O_2$

♦ Learning Exercise 13.1B

Indicate the number of monosaccharide units (one, two, or many) in each of the following carbohydrates:

1. sucrose, a disaccharide _____   2. cellulose, a polysaccharide _____

3. glucose, a monosaccharide _____   4. amylose, a polysaccharide _____

5. maltose, a disaccharide _____

Answers   **1.** two   **2.** many   **3.** one   **4.** many   **5.** two

♦ Learning Exercise 13.1C

Identify the following monosaccharides as aldotrioses, ketotrioses, tetroses, pentoses, or hexoses:

1. CH$_2$OH
    |
    C=O
    |
    CH$_2$OH

2. H   O
     \ //
      C
      |
    H—C—OH
      |
    H—C—OH
      |
    HO—C—H
      |
     CH$_2$OH

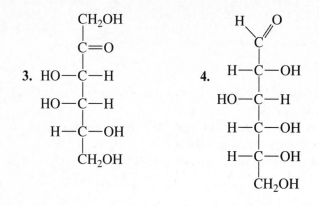

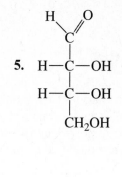

1. _____   2. _____   3. _____

4. _____   5. _____

Answers   1. ketotriose   2. aldopentose   3. ketohexose
             4. aldohexose   5. aldotetrose

13.2 Fischer Projections of Monosaccharides

- In a Fischer projection, the carbon chain is written vertically, with the most oxidized carbon (usually an aldehyde or ketone) at the top.
- In the Fischer projection of a monosaccharide, the chiral —OH *farthest* from the carbonyl group (C=O) is written on the left side in the L isomer and on the right side in the D isomer.
- The carbon atom in the —CH₂OH group at the bottom of the Fischer projection is not chiral because it does not have four different groups bonded to it.
- Important monosaccharides are the aldopentose ribose, the aldohexoses glucose and galactose, and the ketohexose fructose.

MC

Tutorial: Forms of Carbohydrates
Tutorial: Drawing Fischer Projections of Monosaccharides
Tutorial: Identifying D and L Sugars
Tutorial: Identifying Chiral Carbons in Monosaccharides
Tutorial: Drawing Cyclic Sugars

♦ **Learning Exercise 13.2A**

Identify each of the following Fischer projections of sugars as the D or L isomer:

1. 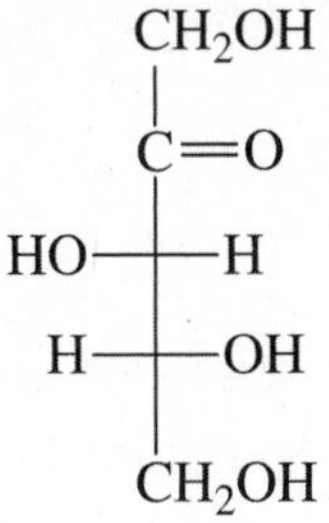
$$\begin{array}{c} CH_2OH \\ | \\ C=O \\ HO\text{---}|\text{---}H \\ H\text{---}|\text{---}OH \\ CH_2OH \end{array}$$
_____ Xylulose

2. 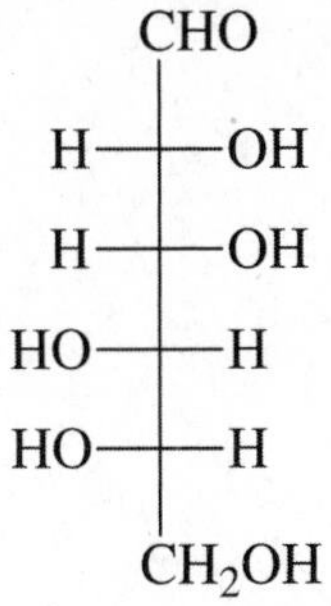
$$\begin{array}{c} CHO \\ H\text{---}|\text{---}OH \\ H\text{---}|\text{---}OH \\ HO\text{---}|\text{---}H \\ HO\text{---}|\text{---}H \\ CH_2OH \end{array}$$
_____ Mannose

3. 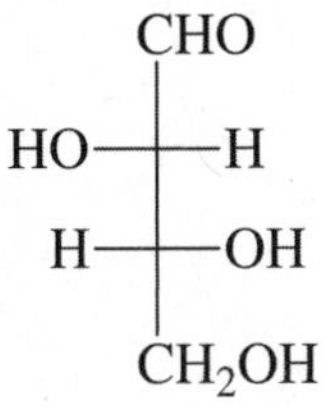
$$\begin{array}{c} CHO \\ HO\text{---}|\text{---}H \\ H\text{---}|\text{---}OH \\ CH_2OH \end{array}$$
_____ Threose

4. 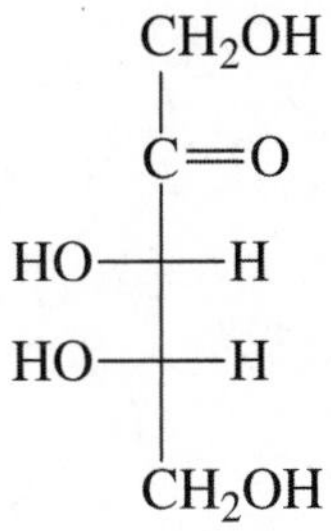
$$\begin{array}{c} CH_2OH \\ | \\ C=O \\ HO\text{---}|\text{---}H \\ HO\text{---}|\text{---}H \\ CH_2OH \end{array}$$
_____ Ribulose

Answers   1. D–Xylulose   2. L–Mannose   3. D–Threose   4. L–Ribulose

♦ **Learning Exercise 13.2B**

Draw the mirror image of each of the sugars in learning exercise 13.2A and give the D or L name.

1.

2.

3.

4.

$$CH_2OH$$
$$C=O$$

1. 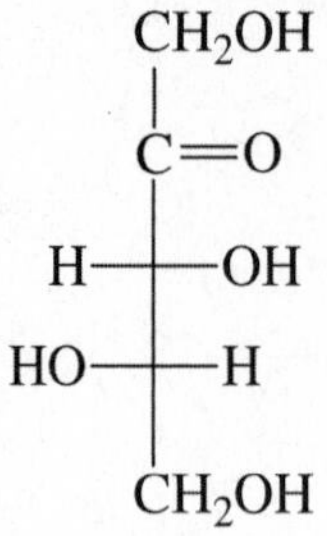   H——OH
     HO——H
     CH_2OH

L-Xylulose

$$CHO$$

2. 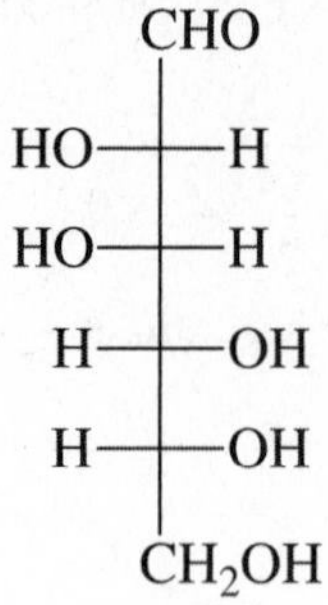   HO——H
     HO——H
     H——OH
     H——OH
     CH_2OH

D-Mannose

$$CHO$$

3. 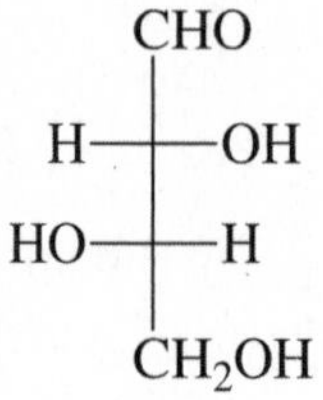   H——OH
     HO——H
     CH_2OH

L-Threose

$$CH_2OH$$
$$C=O$$

4. 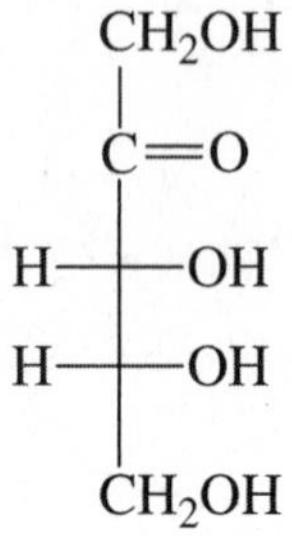   H——OH
     H——OH
     CH_2OH

D-Ribulose

♦ **Learning Exercise 13.2C**

Identify the monosaccharide (glucose, fructose, or galactose) that fits each of the following descriptions:

1. a building block in cellulose   _____

2. also known as fruit sugar   _____

3. accumulates in the disease known as *galactosemia*   _____

4. the most common monosaccharide   _____

5. the sweetest monosaccharide   _____

Answers   1. glucose   2. fructose   3. galactose   4. glucose   5. fructose

♦ **Learning Exercise 13.2D**

Draw the open-chain structure for the following monosaccharides:

D–glucose   D–galactose   D–fructose

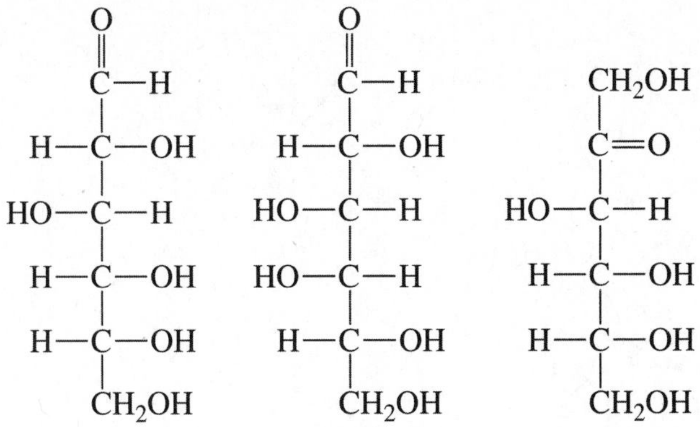

| D-Glucose | D-Galactose | D-Fructose |

13.3 Haworth Structures of Monosaccharides

- The Haworth structure is a representation of the cyclic, stable form of monosaccharides, which has five or six atoms. The cyclic structure forms by a reaction between an —OH on carbon 5 of hexoses and the carbonyl group of the same molecule.
- The formation of a new hydroxyl group on carbon 1 (or 2 in fructose) gives α and β isomers of the cyclic monosaccharide. Because the molecule opens and closes continuously in solution, both α and β isomers are present.

Self Study Activity: Forms of Carbohydrates
Tutorial: Drawing Cyclic Sugars

Guide to Drawing Haworth Structures	
Step 1	Turn the open-chain condensed structural formula clockwise 90°.
Step 2	Fold the chain into a hexagon and bond the O on carbon 5 to the carbonyl group.
Step 3	Draw the new —OH group on carbon 1 down to give the α isomer or up to give the β isomer.

♦ **Learning Exercise 13.3**

Draw the α isomer of the Haworth structure for each of the following:

1. D-glucose   2. D-galactose   3. D-fructose

Answers

1.

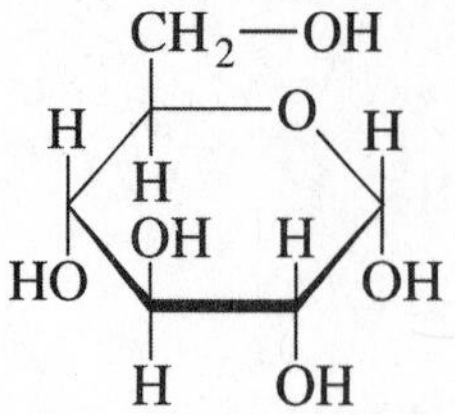

2.

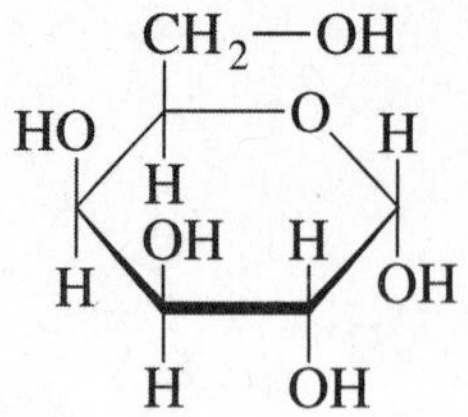

3.

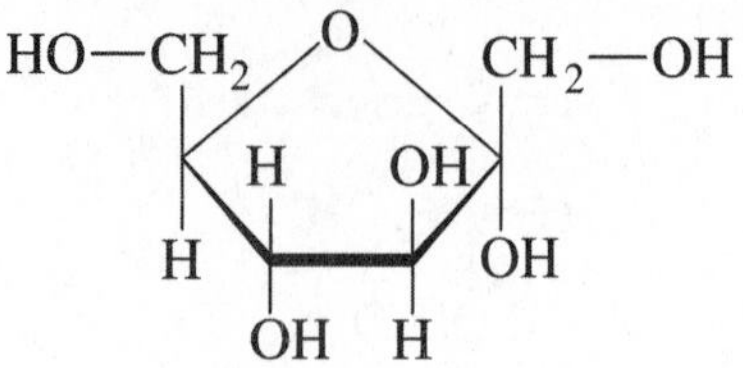

13.4 Chemical Properties of Monosaccharides

- Monosaccharides contain functional groups that undergo oxidation or reduction.
- Monosaccharides are called *reducing sugars* because the aldehyde group (also available in ketoses) is oxidized by a metal ion such as Cu^{2+} in Benedict's solution, which is reduced.
- Monosaccharides are also reduced to give sugar alcohols.

(MC)

Tutorial: Reactions of Monosaccharides
Case Study: Diabetes and Blood Glucose

♦ **Learning Exercise 13.4**

What changes occur when a reducing sugar reacts with Benedict's reagent?

Answer The aldehyde group of the reducing sugar is oxidized to a carboxylic acid group; the Cu^{2+} ion in Benedict's reagent is reduced to Cu^+, which forms a brick-red solid of Cu_2O.

13.5 Disaccharides

- Disaccharides have two monosaccharide units joined together by a glycosidic bond:
    monosaccharide (1) + monosaccharide (2) → disaccharide + H_2O
- In the most common disaccharides, maltose, lactose, and sucrose, there is at least one glucose unit.
- In the disaccharide maltose, two glucose units are linked by an α-1,4 bond.
- When a disaccharide is hydrolyzed by water in the presence of acid or an enzyme, the products are a glucose unit and one other monosaccharide.

$$Maltose + H_2O \rightarrow glucose + glucose$$
$$Lactose + H_2O \rightarrow glucose + galactose$$
$$Sucrose + H_2O \rightarrow glucose + fructose$$

♦ Learning Exercise 13.5

For the following disaccharides, state (a) the monosaccharide units, (b) the type of glycosidic bond, and (c) the name of the disaccharide:

1.

2.

3.

4.

	a. Monosaccharide(s)	b. Type of glycosidic bond	c. Name of disaccharide
1.			
2.			
3.			
4.			

Answers

1. (a) two glucose units   (b) α-1,4-glycosidic bond   (c) β-maltose
2. (a) galactose + glucose   (b) β-1,4-glycosidic bond   (c) α-lactose
3. (a) fructose + glucose   (b) α,β-1,2-glycosidic bond   (c) sucrose
4. (a) two glucose units   (b) α-1,4-glycosidic bond   (c) α-maltose

13.6 Polysaccharides

- Polysaccharides are polymers of monosaccharide units.
- Starches consist of amylose, an unbranched chain of glucose, and amylopectin, a branched polymer of glucose. Glycogen, the storage form of glucose in animals, is similar to amylopectin with more branching.
- Cellulose is also a polymer of glucose, but in cellulose the glycosidic bonds are β bonds rather than α bonds as in the starches. Humans can digest starches to obtain energy, but cannot digest cellulose. However, cellulose is important as a source of fiber in our diets.

Self Study Activity: Forms of Carbohydrates   (MC)

♦ **Learning Exercise 13.6**

List the monosaccharides and describe the glycosidic bonds in each of the following carbohydrates:

	Monosaccharides	Type(s) of glycosidic bonds
a. amylose	_____	_____
b. amylopectin	_____	_____
c. glycogen	_____	_____
d. cellulose	_____	_____

Answers
a. glucose; α-1,4-glycosidic bonds
b. glucose; α-1,4- and α-1,6-glycosidic bonds
c. glucose; α-1,4- and α-1,6-glycosidic bonds
d. glucose; β-1,4-glycosidic bonds

Checklist for Chapter 13

You are ready to take the practice test for Chapter 13. Be sure that you have accomplished the following learning goals for this chapter. If you are not sure, review the section listed at the end of the goal. Then apply your new skills and understanding to the Practice Test.

After studying Chapter 13, I can successfully:

_____ Classify carbohydrates as monosaccharides, disaccharides, and polysaccharides (13.1).

_____ Classify a monosaccharide as aldose or ketose and indicate the number of carbon atoms (13.1).

_____ Draw and identify D and L Fischer projections for carbohydrate molecules (13.2).

_____ Draw the open-chain structures for D-glucose, D-galactose, and D-fructose (13.2).

_____ Draw and identify the cyclic structures of monosaccharides (13.3).

_____ Describe some chemical properties of carbohydrates (13.4).

_____ Describe the monosaccharide units and linkages in disaccharides (13.5).

_____ Describe the structural features of amylose, amylopectin, glycogen, and cellulose (13.6).

Practice Test for Chapter 13

1. The requirements for photosynthesis are
   A. Sun
   B. Sun and water
   C. water and carbon dioxide
   D. Sun, water, and carbon dioxide
   E. carbon dioxide and Sun

2. What are the products of photosynthesis?
   A. carbohydrates
   B. carbohydrates and oxygen
   C. carbon dioxide and oxygen
   D. carbohydrates and carbon dioxide
   E. water and oxygen

3. The name "carbohydrate" came from the fact that
   A. carbohydrates are hydrates of water
   B. carbohydrates contain hydrogen and oxygen in a 2:1 ratio
   C. carbohydrates contain a great quantity of water
   D. all plants produce carbohydrates
   E. carbon and hydrogen atoms are abundant in carbohydrates

4. What functional groups are in the open chains of monosaccharides?
   A. hydroxyl groups
   B. aldehyde groups
   C. ketone groups
   D. hydroxyl and aldehyde or ketone groups
   E. hydroxyl and ether groups

5. What is the classification of the following sugar?

   CH$_2$OH
   |
   C=O
   |
   CH$_2$OH

   A. aldotriose   B. ketotriose   C. aldotetrose   D. ketotetrose   E. ketopentose

For questions 6 through 10, refer to the following monosaccharide:

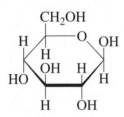

6. It is the cyclic structure of a(n)
   A. aldotriose   B. ketopentose   C. aldopentose   D. aldohexose   E. aldoheptose

7. This is a Haworth structure of
   A. fructose   B. glucose   C. ribose   D. glyceraldehyde   E. galactose

8. It is at least one of the products of the complete hydrolysis of
   A. maltose   B. sucrose   C. lactose   D. glycogen   E. all of these

9. A Benedict's test with this sugar would
    A. be positive   **B.** be negative
    C. produce a blue precipitate   **D.** give no color change
    E. produce a silver mirror

10. It is the monosaccharide unit used to build polymers of
    A. amylose   **B.** amylopectin   **C.** cellulose   **D.** glycogen   **E.** all of these

For questions 11 through 15, identify each carbohydrate described as one of the following:
    A. maltose   **B.** sucrose   **C.** cellulose   **D.** amylopectin   **E.** glycogen

11. _____ a disaccharide that is not a reducing sugar

12. _____ a disaccharide that occurs as a breakdown product of amylose

13. _____ a carbohydrate that is produced as a storage form of energy in plants

14. _____ the storage form of energy in humans

15. _____ a carbohydrate that is used for structural purposes by plants

For questions 16 through 20, select answers from the following:
    A. amylose   **B.** cellulose   **C.** glycogen   **D.** lactose   **E.** glucose

16. _____ a polysaccharide composed of many glucose units linked by α-1,4-glycosidic bonds

17. _____ a sugar containing both glucose and galactose

18. _____ a sugar composed of glucose units joined by both α-1,4- and α-1,6-glycosidic bonds

19. _____ a monosaccharide

20. _____ a carbohydrate composed of glucose units joined by β-1,4-glycosidic bonds

For questions 21 through 25, select answers from the following:
    A. glucose   **B.** lactose   **C.** sucrose   **D.** maltose

21. _____ a sugar composed of glucose and fructose

22. _____ also called table sugar

23. _____ found in milk and milk products

24. _____ gives sorbitol upon reduction

25. _____ gives galactose upon hydrolysis

Answers to the Practice Test

1. D	**2.** B	**3.** B	**4.** D	**5.** B
6. D	**7.** B	**8.** E	**9.** A	**10.** E
11. B	**12.** A	**13.** D	**14.** E	**15.** C
16. A	**17.** D	**18.** C	**19.** E	**20.** B
21. C	**22.** C	**23.** B	**24.** A	**25.** B

14
Carboxylic Acids, Esters, Amines, and Amides

Study Goals

- Name and draw condensed structural and skeletal formulas of carboxylic acids and esters.
- Describe the solubility of carboxylic acids.
- Write equations for the ionization of carboxylic acids in water. Write an equation for esterification.
- Write equations for the hydrolysis and saponification of esters.
- Name and draw condensed structural and skeletal formulas of amines and amides.
- Describe the ionization of amines in water.
- Describe the solubility of amines and amides in water.
- Write equations for the neutralization and amidation of amines.
- Write equations for the acid and base hydrolysis of amides.

Think About It

1. Why do vinegar and citrus juices taste sour?

2. What type of compound gives flowers and fruits their pleasant aromas?

3. Fish smell "fishy," but lemon juice removes the "fishy" odor. Why?

4. What functional groups are often found in tranquilizers and hallucinogens?

Key Terms

Match the key term with the correct statement shown below.

  a. carboxylic acid   **b.** saponification   **c.** esterification   **d.** hydrolysis
  e. ester   **f.** amine   **g.** amide   **h.** alkaloid

1. _____ an organic compound containing the carboxyl group (—COOH)

2. _____ a reaction of a carboxylic acid and an alcohol in the presence of an acid catalyst

3. _____ a type of organic compound that produces pleasant aromas in flowers and fruits

4. _____ the hydrolysis of an ester with a strong base producing a salt of the carboxylic acid and an alcohol

5. _____ the splitting of a molecule such as an ester by the addition of water in the presence of an acid

6. _____ a nitrogen-containing compound that is produced by plants and is active physiologically

7. _____ the hydrolysis of this compound produces a carboxylic acid and an amine

8. _____ an organic compound that contains an —NH_2 group

Answers	**1.** a	**2.** c	**3.** e	**4.** b
	5. d	**6.** h	**7.** g	**8.** f

14.1 Carboxylic Acids

- In the IUPAC system, a carboxylic acid is named by replacing the *ane* ending with *oic acid*. Simple acids usually are named by the common naming system using the prefixes: *form* (1C), *acet* (2C), *propion* (3C), and *butyr* (4C), followed by *ic acid*.

H—C—OH   CH$_3$—C—OH   CH$_3$—CH$_2$—CH$_2$—C—OH

Methanoic acid   Ethanoic acid   Butanoic acid   Pentanoic acid
(formic acid)   (acetic acid)   (butyric acid)

- The aromatic carboxylic acid, which is named benzoic acid, has a carboxyl group attached to carbon 1, and substituents numbered in the direction that gives the smallest numbers.

C—OH

Benzonic acid

MC	
Self Study Activity: Carboxylic Acids	
Tutorial: Naming and Drawing Carboxylic Acids	

Guide to Naming Carboxylic Acids	
Step 1	Identify the carbon chain containing the carboxyl group and replace the *e* in the alkane name with *oic acid*.
Step 2	Give the location and name of each substituent on the main chain by counting the carboxyl carbon as 1.

♦ Learning Exercise 14.1A

Give the IUPAC name for each of the following carboxylic acids:

1. $\mathrm{CH_3-\overset{O}{\overset{\|}{C}}-OH}$

2. $\mathrm{CH_3-\overset{OH}{\overset{|}{C}H}-\overset{O}{\overset{\|}{C}}-OH}$

3. $\mathrm{CH_3-\overset{CH_3}{\overset{|}{C}H}-CH_2-\overset{O}{\overset{\|}{C}}-OH}$

4. 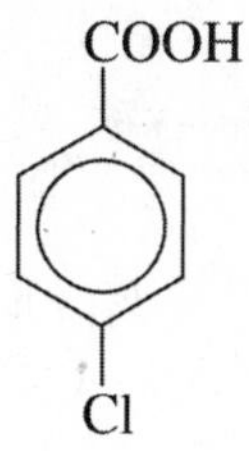

Answers
1. ethanoic acid
2. 2-hydroxypropanoic acid
3. 3-methylbutanoic acid
4. 4-chlorobenzoic acid

♦ Learning Exercise 14.1B

A. Draw the condensed structural formulas for **1.** through **3.** and the skeletal formulas for **4.** through **6.**

1. acetic acid
2. 3-hydroxypropanoic acid
3. formic acid
4. 2-bromobutanoic acid
5. benzoic acid
6. 3-methylpentanoic acid

Answers

1. $\mathrm{CH_3-\overset{O}{\overset{\|}{C}}-OH}$

2. $\mathrm{HO-CH_2-CH_2-\overset{O}{\overset{\|}{C}}-OH}$

3. $\mathrm{H-\overset{O}{\overset{\|}{C}}-OH}$

4. 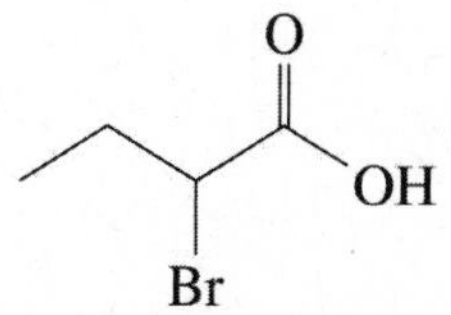

5. 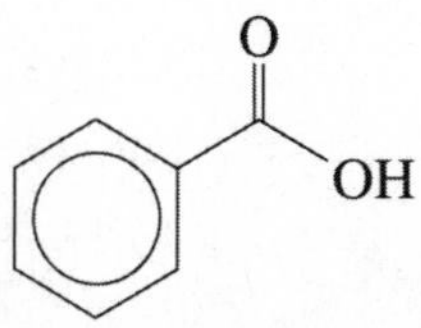

6. 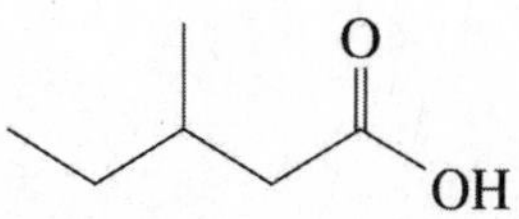

14.2 Properties of Carboxylic Acids

- Carboxylic acids with one to four carbon atoms are very soluble in water.
- As weak acids, carboxylic acids ionize slightly in water to form acidic solutions of H_3O^+ and a carboxylate ion.
- When bases neutralize carboxylic acids, the products are carboxylate salts and water.

Tutorial: Properties of Carboxylic Acids (MC)

◆ Learning Exercise 14.2A

Indicate whether each of the following carboxylic acids is soluble in water:

1. _____ hexanoic acid   2. _____ propanoic acid   3. _____ formic acid

4. _____ acetic acid   5. _____ benzoic acid   6. _____ octanoic acid

Answers   **1.** no   **2.** yes   **3.** yes   **4.** yes   **5.** no   **6.** no

◆ Learning Exercise 14.2B

Draw the condensed structural formulas for the products for the ionization of the following carboxylic acids in water:

1.
$$CH_3-CH_2-\overset{\overset{\displaystyle O}{\|}}{C}-OH + H_2O \rightleftharpoons$$

2. benzoic acid + $H_2O \rightleftharpoons$

Answers   1.
$$CH_3-CH_2-\overset{\overset{\displaystyle O}{\|}}{C}-O^- + H_3O^+$$
2. 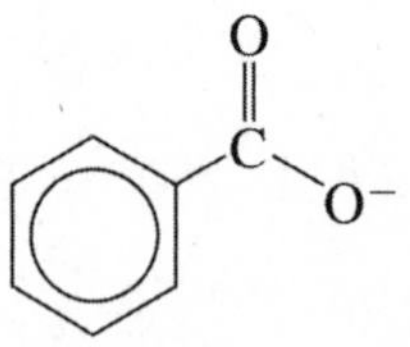
$$+ H_3O^+$$

♦ **Learning Exercise 14.2C**

Draw the condensed structural formulas for the products and names for each of the following:

1. $CH_3-CH_2-\overset{\overset{\displaystyle O}{\|}}{C}-OH + NaOH \longrightarrow$

2. formic acid + KOH $\longrightarrow$

Answers   1. $CH_3-CH_2-\overset{\overset{\displaystyle O}{\|}}{C}-O^-Na^+ + H_2O$ 2. $H-\overset{\overset{\displaystyle O}{\|}}{C}-O^-K^+ + H_2O$

Sodium propanoate
(sodium propionate)

Potassium
methanoate
(potassium formate)

14.3 Esters

- In the presence of a strong acid, carboxylic acids react with alcohols to produce esters and water.
- The names of esters consist of two words, one from the alcohol and the other from the carboxylic acid with the *ic* ending replaced by *ate*.

$CH_3-\overset{\overset{\displaystyle O}{\|}}{C}-O-CH_3$ methyl ethanoate (IUPAC) or methyl acetate (common)

- In hydrolysis, esters are split apart by a reaction with water. When the catalyst is an acid, the products are a carboxylic acid and an alcohol.

$CH_3-\overset{\overset{\displaystyle O}{\|}}{C}-O-CH_3 + H_2O \xrightarrow{H^+} CH_3-\overset{\overset{\displaystyle O}{\|}}{C}-OH + HO-CH_3$

Methyl acetate   Acetic acid   Methyl alcohol

- The hydrolysis of an ester in the presence of a base produces a carboxylate salt and an alcohol.

$CH_3-\overset{\overset{\displaystyle O}{\|}}{C}-O-CH_3 + NaOH \longrightarrow CH_3-\overset{\overset{\displaystyle O}{\|}}{C}-O^-Na^+ + HO-CH_3$

Methyl acetate   Sodium acetate   Methyl alcohol

- In saponification, long-chain fatty acids from fats react with strong bases to produce salts of the fatty acids, which are soaps.

(MC)

Tutorial: Writing Esterification Equations
Tutorial: Formation of Esters from Carboxylic Acids

Guide to Naming Esters	
Step 1	Write the name of carbon chain from the alcohol as an *alkyl* group.
Step 2	Change the *ic acid* of the acid name to *ate*.

♦ **Learning Exercise 14.3A**

Draw the condensed structural formulas for the products of the following reactions:

1. CH_3—$\overset{\overset{O}{\|}}{C}$—$OH$ + CH_3—OH $\xrightarrow{H^+}$

2. H—$\overset{\overset{O}{\|}}{C}$—$OH$ + CH_3—CH_2—OH $\xrightarrow{H^+}$

3. 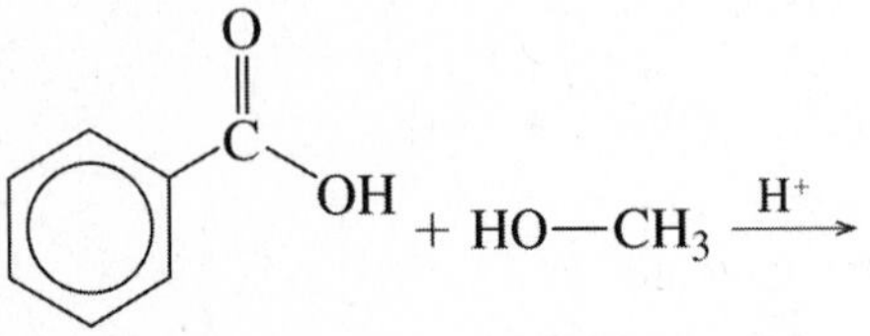 + HO—CH_3 $\xrightarrow{H^+}$

4. propanoic acid and ethanol $\xrightarrow{H^+}$

Answers

1. CH_3—$\overset{\overset{O}{\|}}{C}$—$O$—$CH_3$ + H_2O

2. H—$\overset{\overset{O}{\|}}{C}$—$O$—$CH_2$—$CH_3$ + H_2O

3. 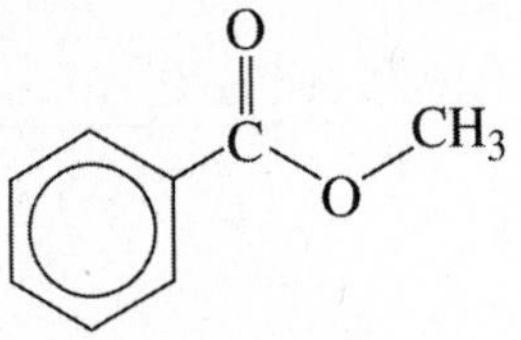 + H_2O

4. CH_3—CH_2—$\overset{\overset{O}{\|}}{C}$—$O$—$CH_2$—$CH_3$ + H_2O

Tutorial: Naming Esters

♦ **Learning Exercise 14.3B**

Name each of the following esters:

1. 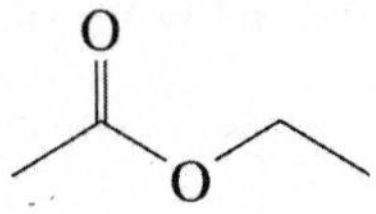 $CH_3-C(=O)-O-CH_2-CH_3$ (skeletal)

2. $CH_3-CH_2-CH_2-C(=O)-O-CH_3$

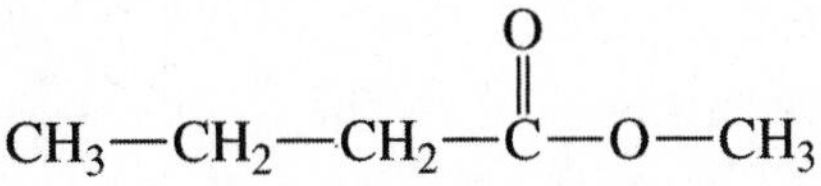

3. $CH_3-CH_2-C(=O)-O-CH_2-CH_2-CH_3$

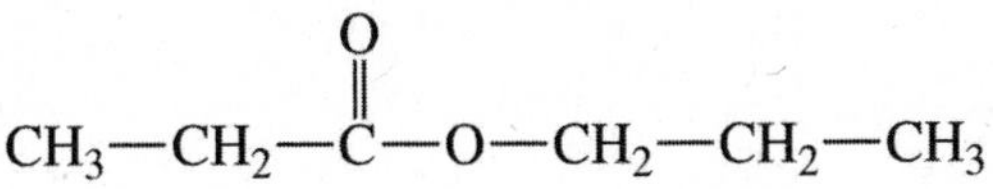

4. 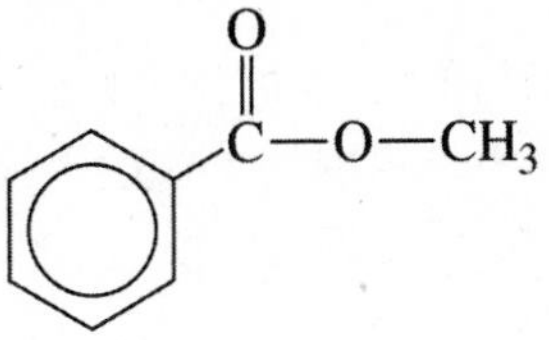 $C_6H_5-C(=O)-O-CH_3$ (benzene ring)

Answers
1. ethyl ethanoate (ethyl acetate)
2. methyl butanoate (methyl butyrate)
3. propyl propanoate (propyl propionate)
4. methyl benzoate

♦ **Learning Exercise 14.3C**

Draw the condensed structural formulas for 1 through 3 and the skeletal formula for 4.

1. propyl acetate

2. ethyl butyrate

3. ethyl benzoate

4. ethyl propanoate

Answers

1. 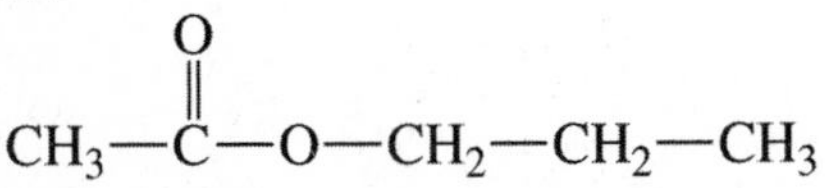 $CH_3-C(=O)-O-CH_2-CH_2-CH_3$

2. $CH_3-CH_2-CH_2-C(=O)-O-CH_2-CH_3$

3. $C_6H_5-C(=O)-O-CH_2-CH_3$ (benzene ring)

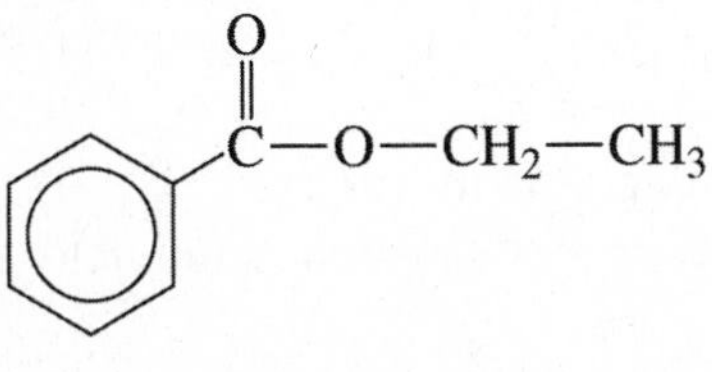

4. 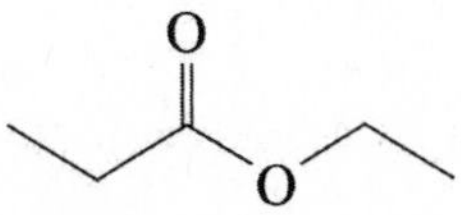 skeletal formula of ethyl propanoate

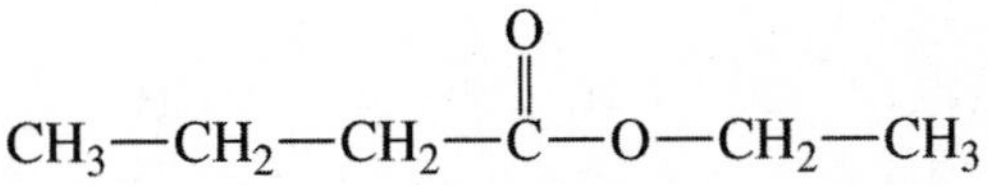

Tutorial: Hydrolysis of Esters

♦ **Learning Exercise 14.3D**

Draw the condensed structural formulas for the products of hydrolysis or saponification for the following esters:

$$
1.\ CH_3-CH_2-CH_2-\overset{\overset{\displaystyle O}{\|}}{C}-O-CH_3 + H_2O \xrightarrow{\ H^+\ }
$$

$$
2.\ CH_3-\overset{\overset{\displaystyle O}{\|}}{C}-O-CH_3 + NaOH \longrightarrow
$$

$$
3.\ CH_3-CH_2-\overset{\overset{\displaystyle O}{\|}}{C}-O-CH_2-CH_3 + KOH \longrightarrow
$$

$$
4.\ CH_3-CH_2-\overset{\overset{\displaystyle O}{\|}}{C}-O-CH_3 + H_2O \xrightarrow{\ H^+\ }
$$

Answers

$$
1.\ CH_3-CH_2-CH_2-\overset{\overset{\displaystyle O}{\|}}{C}-OH + HO-CH_3 \qquad 2.\ CH_3-\overset{\overset{\displaystyle O}{\|}}{C}-O^-Na^+ + CH_3-OH
$$

$$
3.\ CH_3-CH_2-\overset{\overset{\displaystyle O}{\|}}{C}-O^-K^+ + CH_3-CH_2-OH \qquad 4.\ CH_3-CH_2-\overset{\overset{\displaystyle O}{\|}}{C}-OH + HO-CH_3
$$

(MC)

Tutorial: Name that Amine

Tutorial: Drawing Amines

Self Study Activity: Amine and Amide Functional Groups

14.4 Amines

- Amines are derivatives of ammonia (NH_3), in which alkyl or aromatic groups replace one or more hydrogen atoms.
- Amines are classified as primary, secondary, or tertiary when the nitrogen atom is bonded to one, two, or three alkyl or aromatic groups.

$$
CH_3-NH_2 \qquad CH_3-\overset{\overset{\displaystyle CH_3}{|}}{N}-H \qquad CH_3-\overset{\overset{\displaystyle CH_3}{|}}{N}-CH_3
$$

Primary (1°)   Secondary (2°)   Tertiary (3°)   Primary (1°)

- The N—H bonds in primary and secondary amines form hydrogen bonds.
- Hydrogen bonding allows amines with up to six carbon atoms to be soluble in water.
- In water, amines act as weak bases by accepting protons from water to produce ammonium and hydroxide ions.

$$
CH_3-NH_2 + H_2O \rightleftarrows CH_3-NH_3^+ + OH^-
$$

Methylamine   Methylammonium ion and hydroxide ion

- Strong acids neutralize amines to yield ammonium salts.

$$CH_3-NH_2 + HCl \rightarrow CH_3-NH_3^+Cl^-$$

  Methylamine   Methylammonium chloride

- A heterocyclic amine is a cyclic compound containing one or more nitrogen atoms in the ring.
- Most heterocyclic amines contain five or six atoms in the ring.

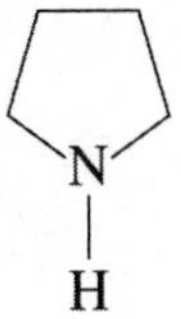

                                               Pyridine

    Pyrrolidine   Pyrrole   Piperidine

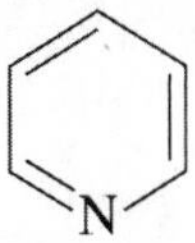

- An alkaloid is a physiologically active amine obtained from plants.

Tutorial: Identifying Types of Heterocyclic Amines

♦ **Learning Exercise 14.4A**

Classify each of the following as a primary (1°), secondary (2°), or tertiary (3°) amine:

1. _____ $CH_3-\overset{\overset{H}{|}}{N}-CH_2-CH_3$

2. _____ (structure shown)

3. _____ (cyclohexyl with NH_2)

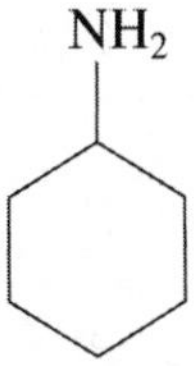

4. _____ (benzene ring with NH_2 and CH_3)

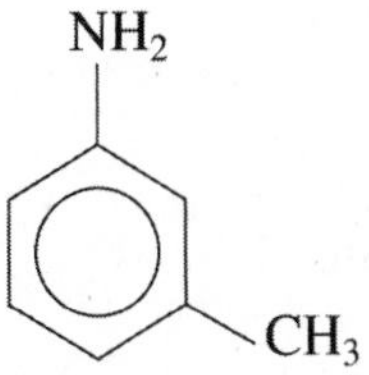

5. _____ $CH_3-CH_2-\overset{\overset{CH_3}{|}}{N}-CH_3$

Answers   **1.** 2°   **2.** 2°   **3.** 1°   **4.** 1°   **5.** 3°

♦ **Learning Exercise 14.4B**

Name each of the amines in problem 14.4A.

1. _____   2. _____

3. _____   4. _____

5. _____

Answers   **1.** ethylmethylamine   **2.** butylethylamine   **3.** cyclohexylamine
                 4. 3-methylaniline   **5.** ethyldimethylamine

♦ **Learning Exercise 14.4C**

Draw the condensed structural formula for each of the following amines:

1. isopropylamine

2. butylethylmethylamine

3. 3-bromoaniline

Answers

1.
$$CH_3\overset{\overset{\displaystyle NH_2}{|}}{-}CH-CH_3$$

2.
$$CH_3-CH_2\overset{\overset{\displaystyle CH_3}{|}}{-}N-CH_2-CH_2-CH_2-CH_3$$

3. 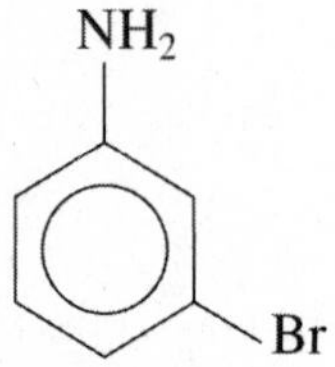 Br

---

Tutorial: Reactions of Amines
Case Study: Death by Chocolate?

MC

♦ **Learning Exercise 14.4D**

Draw the condensed structural formula for the products for each of the following reactions:

1. $CH_3-CH_2-NH_2 + H_2O \rightleftarrows$

2. $CH_3-CH_2-CH_2-NH_2 + HCl \rightarrow$

3. $CH_3CH_2-NH-CH_3 + HCl \rightarrow$

4.

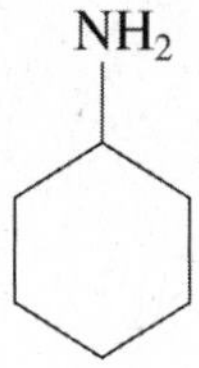

$$\text{(cyclohexyl-}NH_2) + HBr \longrightarrow$$

5.

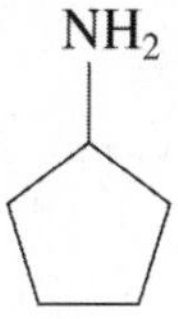

$$\text{(cyclopentyl-}NH_2) + H_2O \rightleftharpoons$$

6. $CH_3\!-\!CH_2\!-\!NH_3{}^+Cl^- + NaOH \rightarrow$

Answers

1. $CH_3\!-\!CH_2\!-\!NH_3{}^+ + OH^-$ **2.** $CH_3\!-\!CH_2\!-\!CH_2\!-\!NH_3{}^+Cl^-$ **3.** $CH_3\!-\!CH_2\!-\!\overset{+}{N}H_2\!-\!CH_3Cl^-$

4. 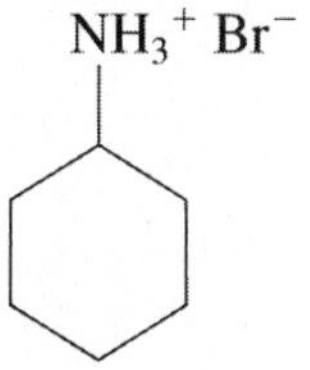 $NH_3{}^+\,Br^-$

5. 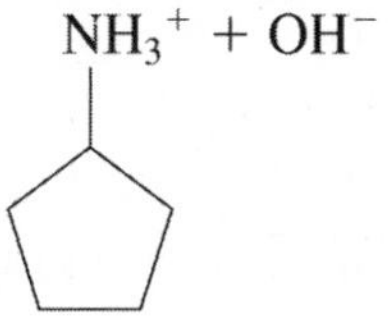 $NH_3{}^+ + OH^-$

6. $CH_3\!-\!CH_2\!-\!NH_2 + NaCl + H_2O$

♦ **Learning Exercise 14.4E**

Match each of following heterocyclic structures with the correct name:

a. (pyrrole structure)   **b.** 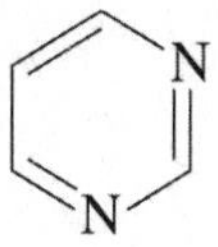 (pyrimidine structure)   **c.** 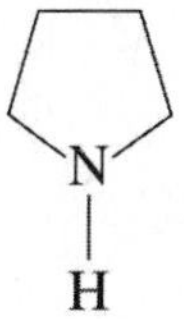 (pyrrolidine structure)

d. 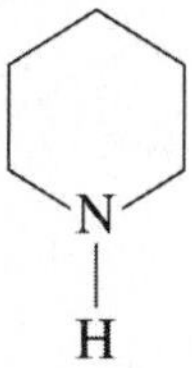 (piperidine structure)   **e.** 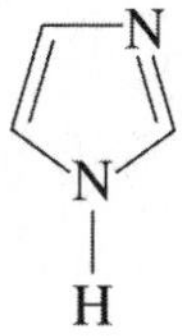 (imidazole structure)   **f.** 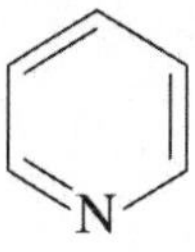 (pyridine structure)

1. _____ pyrrolidine   **2.** _____ imidazole   **3.** _____ pyridine
4. _____ pyrrole   **5.** _____ pyrimidine   **6.** _____ piperidine

Answers   **1.** c   **2.** e   **3.** f   **4.** a   **5.** b   **6.** d

---

(MC)

Self Study Activity: Amine and Amide Functional Groups

14.5 Amides

- Amides are derivatives of carboxylic acids in which an amine group replaces the —OH group in the carboxylic acid.
- Amides are named by replacing the *ic acid* or *oic acid* ending by *amide*.

$$CH_3-\overset{\displaystyle O}{\overset{\|}{C}}-NH_2 \qquad \text{Ethanamide (acetamide)}$$

- Amides undergo acid and base hydrolysis to produce the carboxylic acid (or carboxylate salt) and the amine (or amine salt).

$$CH_3-\overset{\displaystyle O}{\overset{\|}{C}}-NH_2 + HCl + H_2O \longrightarrow CH_3-\overset{\displaystyle O}{\overset{\|}{C}}-OH + NH_4^+\ Cl^-$$

$$CH_3-\overset{\displaystyle O}{\overset{\|}{C}}-NH_2 + NaOH \longrightarrow CH_3-\overset{\displaystyle O}{\overset{\|}{C}}-O^-Na^+ + NH_3$$

Tutorial: Amidation Reactions

♦ **Learning Exercise 14.5A**

Draw the condensed structural formula for the amide formed in each of the following reactions:

1. $CH_3-CH_2-\overset{\displaystyle O}{\overset{\|}{C}}-OH + NH_3 \xrightarrow{\text{Heat}}$

2. 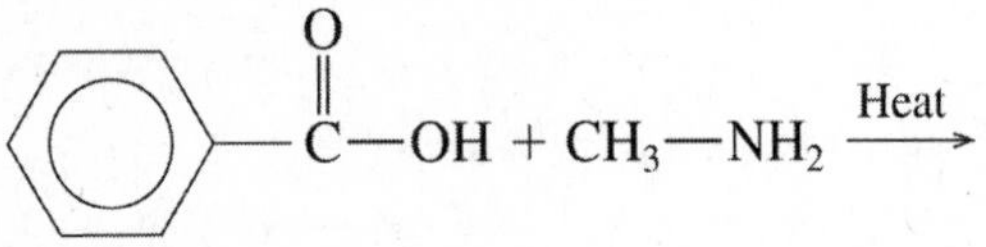 $\text{C}_6\text{H}_5-\overset{\displaystyle O}{\overset{\|}{C}}-OH + CH_3-NH_2 \xrightarrow{\text{Heat}}$

3. $CH_3-\overset{\displaystyle O}{\overset{\|}{C}}-OH + \overset{\displaystyle CH_3}{\overset{|}{NH}}-CH_3 \xrightarrow{\text{Heat}}$

Answers

1. $CH_3-CH_2-\overset{\displaystyle O}{\overset{\|}{C}}-NH_2$

2. 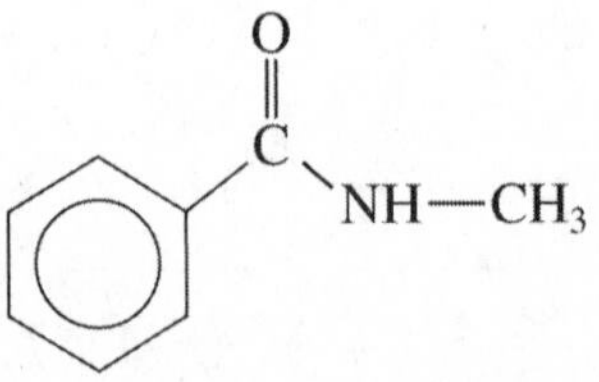 benzamide N-methyl

3. $CH_3-\overset{\displaystyle O}{\overset{\|}{C}}-\overset{\displaystyle CH_3}{\overset{|}{N}}-CH_3$

♦ **Learning Exercise 14.5B**

Name the following amides:

1. $CH_3-CH_2-\overset{\overset{\displaystyle O}{\|}}{C}-NH_2$ _____

2.

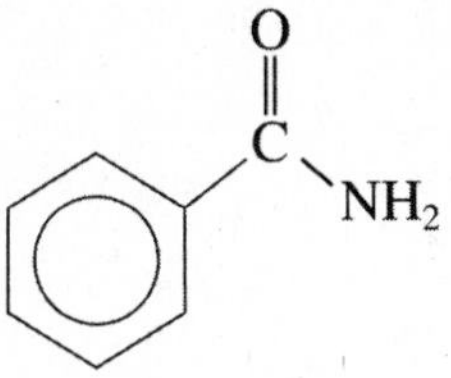

$\overset{\overset{\displaystyle O}{\|}}{C}\diagdown NH_2$ attached to benzene ring _____

3. $CH_3-CH_2-CH_2-CH_2-\overset{\overset{\displaystyle O}{\|}}{C}-NH_2$ _____

4. $CH_3-\overset{\overset{\displaystyle O}{\|}}{C}-NH_2$ _____

Answers   1. propanamide (propionamide)   **2.** benzamide
              3. pentanamide   **4.** ethanamide (acetamide)

♦ **Learning Exercise 14.5C**

Draw the condensed structural formula for each of the following amides:

1. methanamide   **2.** butanamide

3. 3-chloropentanamide   **4.** benzamide

Answers

1. $H-\overset{\overset{\displaystyle O}{\|}}{C}-NH_2$

2. $CH_3-CH_2-CH_2-\overset{\overset{\displaystyle O}{\|}}{C}-NH_2$

3. $CH_3-CH_2-\overset{\overset{\displaystyle Cl}{|}}{CH}-CH_2-\overset{\overset{\displaystyle O}{\|}}{C}-NH_2$

4. 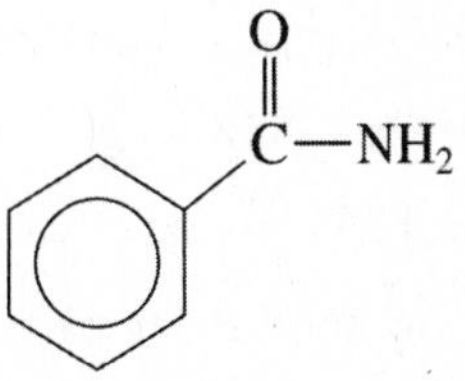 benzene ring with $\overset{\overset{\displaystyle O}{\|}}{C}-NH_2$

Tutorial: Hydrolysis of Amides

♦ **Learning Exercise 14.5D**

Using condensed structural formulas, draw the products for the hydrolysis of each of the following with HCl and NaOH:

1. $CH_3-CH_2-\overset{\overset{\displaystyle O}{\|}}{C}-NH_2$

2. $CH_3-\overset{\overset{\displaystyle O}{\|}}{C}-NH_2$

Answers

1. (HCl) $CH_3-CH_2-\overset{\overset{\displaystyle O}{\|}}{C}-OH + NH_4^+Cl^-$ (NaOH) $CH_3-CH_2-\overset{\overset{\displaystyle O}{\|}}{C}-O^-Na^+ + NH_3$

2. (HCl) $CH_3-\overset{\overset{\displaystyle O}{\|}}{C}-OH + NH_4^+Cl^-$ (NaOH) $CH_3-\overset{\overset{\displaystyle O}{\|}}{C}-O^-Na^+ + NH_3$

Checklist for Chapter 14

You are ready to take the Practice Test for Chapter 14. Be sure that you have accomplished the following learning goals for this chapter. If you are not sure, review the section listed at the end of the goal. Then apply your new skills and understanding to the Practice Test.

After studying Chapter 14, I can successfully:

_____ Write the IUPAC and common names, and draw condensed structural formulas, of carboxylic acids (14.1).

_____ Describe the solubility and ionization of carboxylic acids in water (14.2).

_____ Describe the behavior of carboxylic acids as weak acids and write the structural formulas for the

products of neutralization (14.2).

_____ Write equations for the preparation, hydrolysis, and saponification of esters (14.3).

_____ Write the IUPAC or common names and draw the condensed structural formulas of esters (14.3).

_____ Classify amines as primary, secondary, or tertiary (14.4).

_____ Write the IUPAC and common names of amines and draw their condensed structural formulas (14.4).

_____ Write equations for the ionization and neutralization of amines (14.4).

_____ Write the IUPAC and common names of amides and draw their condensed structural formulas (14.5).

_____ Write equations for the hydrolysis of amides (14.5).

Practice Test for Chapter 14

Match structures 1 through 5 with functional groups A to E.

    A. alcohol   **B.** aldehyde   **C.** carboxylic acid   **D.** ester   **E.** ketone

1. _____ $CH_3-\overset{\overset{\displaystyle CH_3}{|}}{CH}-CH_2-OH$
    2. _____ $CH_3-CH_2-\overset{\overset{\displaystyle O}{||}}{C}-OH$

3. _____ $CH_3-CH_2-\overset{\overset{\displaystyle O}{||}}{C}-H$
    4. _____ $CH_3-\overset{\overset{\displaystyle O}{||}}{C}-O-CH_3$

5. _____ $CH_3-CH_2-\overset{\overset{\displaystyle O}{||}}{C}-CH_3$

For questions 6 through 10, match each name with the correct structure A to E.

A. $CH_3-\overset{\overset{\displaystyle O}{||}}{C}-O-CH_2-CH_3$ **B.** $CH_3-CH_2-CH_2-\overset{\overset{\displaystyle O}{||}}{C}-O^-Na^+$ **C.** $CH_3-\overset{\overset{\displaystyle O}{||}}{C}-O^-Na^+$

D. $CH_3-CH_2-\overset{\overset{\displaystyle CH_3}{|}}{CH}-\overset{\overset{\displaystyle O}{||}}{C}-OH$ **E.** $CH_3-CH_2-\overset{\overset{\displaystyle O}{||}}{C}-O-CH_3$

6. _____ 2-methylbutanoic acid

7. _____ methyl propanoate

8. _____ sodium butanoate

9. _____ ethyl acetate

10. _____ sodium acetate

11. What is the product when a carboxylic acid reacts with sodium hydroxide?
    A. carboxylate salt   **B.** alcohol   **C.** ester   **D.** aldehyde   **E.** no reaction

12. Carboxylic acids are water soluble due to their
    A. nonpolar nature   **B.** ionic bonds   **C.** ability to lower pH
    D. ability to hydrogen bond   **E.** high melting points

Chapter 14

For questions 13 through 16, refer to the following reactions:

A. $CH_3-\overset{O}{\overset{\|}{C}}-OH + CH_3-OH \xrightarrow{H^+} CH_3-\overset{O}{\overset{\|}{C}}-O-CH_3 + H_2O$

B. $CH_3-\overset{O}{\overset{\|}{C}}-OH + NaOH \longrightarrow CH_3-\overset{O}{\overset{\|}{C}}-O^-Na^+ + H_2O$

C. $CH_3-\overset{O}{\overset{\|}{C}}-O-CH_3 + H_2O \xrightarrow{H^+} CH_3-\overset{O}{\overset{\|}{C}}-OH + CH_3-OH$

D. $CH_3-\overset{O}{\overset{\|}{C}}-O-CH_3 + NaOH \longrightarrow CH_3-\overset{O}{\overset{\|}{C}}-O^-Na^+ + CH_3-OH$

13. _____ is an acid hydrolysis of an ester

14. _____ is a neutralization

15. _____ is a base hydrolysis of an ester

16. _____ is an esterification

17. What is the name of the organic product of reaction A?
    A. methyl acetate   B. acetic acid   C. methyl alcohol
    D. acetaldehyde   E. ethyl methanoate

18. What is the IUPAC name of the following compound?

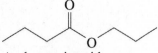

    A. butanoic acid   B. butyl propionate   C. butyl propanoate
    D. propyl butanoate   E. propyl butyrate

19. The ester produced from the reactions of 1-butanol and propanoic acid is
    A. butyl propanoate   B. butyl propanone   C. propyl butyrate
    D. propyl butanone   E. heptanoate

20. The reaction of methyl acetate with NaOH produces
    A. ethanol and formic acid   B. ethanol and sodium formate
    C. ethanol and sodium ethanoate   D. methanol and acetic acid
    E. methanol and sodium acetate

21. Identify the carboxylic acid and alcohol needed to produce

$$CH_3-CH_2-CH_2-\overset{O}{\overset{\|}{C}}-O-CH_2-CH_3$$

    A. propanoic acid and ethanol   B. acetic acid and 1-pentanol   C. acetic acid and 1-butanol
    D. butanoic acid and ethanol   E. hexanoic acid and methanol

22. The name of $CH_3-CH_2-\overset{\overset{\displaystyle O}{\|}}{C}-O-CH_2-CH_3$ is
    - **A.** ethyl acetate
    - **B.** ethyl ethanoate
    - **C.** ethyl propanoate
    - **D.** propyl ethanoate
    - **E.** ethyl butyrate

23. Soaps are
    - **A.** long-chain fatty acids
    - **B.** fatty acid salts
    - **C.** esters of acetic acid
    - **D.** alcohols with 10 carbon atoms
    - **E.** aromatic compounds

24. In a hydrolysis reaction,
    - **A.** an acid reacts with an alcohol
    - **B.** an ester reacts with NaOH
    - **C.** an ester reacts with H_2O
    - **D.** an acid neutralizes a base
    - **E.** water is added to an alkene

25. Esters
    - **A.** have pleasant odors
    - **B.** can undergo hydrolysis
    - **C.** are formed from alcohols and carboxylic acids
    - **D.** have IUPAC names ending in *oate*
    - **E.** all of the above

26. The products of $H-\overset{\overset{\displaystyle O}{\|}}{C}-OH + H_2O$ are
    - **A.** $H-\overset{\overset{\displaystyle O}{\|}}{C}-O^- + H_3O^+$
    - **B.** $H-\overset{\overset{\displaystyle O}{\|}}{C}-\overset{+}{O}H_2 + OH^-$
    - **C.** $H-\overset{\overset{\displaystyle O}{\|}}{C}-O-CH_3$
    - **D.** $CH_3-\overset{\overset{\displaystyle O}{\|}}{C}-OH$
    - **E.** $H-\overset{\overset{\displaystyle O}{\|}}{C}-O^-Na^+ + H_2O$

For questions 27 through 30, classify each amine as A, B, or C.

    A. primary amine   **B.** secondary amine   **C.** tertiary amine

27. _____ $CH_3-\overset{\overset{\displaystyle CH_3}{|}}{CH}-NH_2$

28. _____ $CH_3-CH_2-\overset{\overset{\displaystyle CH_3}{|}}{N}-CH_3$

29. _____ $CH_3-CH_2-\overset{\overset{\displaystyle NH_2}{|}}{CH}-CH_2-CH_3$

30. _____ $CH_3-\overset{\overset{\displaystyle H}{|}}{N}-CH_2-CH_3$

For questions 31 through 33, match the condensed structural formula with the name A to E.

    - **A.** ethyl dimethyl amine
    - **B.** butanamide
    - **C.** *N*-methylacetamide
    - **D.** benzamide
    - **E.** *N*-ethylbutyramide

31. $CH_3-CH_2-\underset{\underset{CH_3}{|}}{N}-CH_3$

32. $CH_3-CH_2-CH_2-\overset{\overset{O}{\parallel}}{C}-NH_2$

33. (benzamide structure: benzene ring—$\overset{\overset{O}{\parallel}}{C}-NH_2$)

For questions 34 through 37, complete the reactions with the correct products A to D.

A. $CH_3-CH_2-\overset{\overset{O}{\parallel}}{C}-OH + NH_4^+Cl^-$ **B.** $CH_3-CH_2-NH_3^+Cl^-$

C. $CH_3-CH_2-CH_2-NH_3^+ + OH^-$ **D.** $CH_3-\overset{\overset{O}{\parallel}}{C}-NH_2$

34. _____ ionization of propylamine in water

35. _____ hydrolysis of propanamide with hydrochloric acid

36. _____ reaction of ethylamine and hydrochloric acid

37. _____ amidation of acetic acid

38. _____ Amines used in drugs are converted to their amine salt because the salt is
A. a solid at room temperature   **B.** soluble in water   **C.** odorless
D. soluble in body fluids   **E.** all of these

For questions 39 through 42, match the name of the alkaloid with its source
A. caffeine   **B.** nicotine   **C.** morphine   **D.** quinine

39. a painkiller from the Oriental poppy plant

40. obtained from the bark of the cinchona tree and used in the treatment of malaria

41. a stimulant obtained from the leaves of tobacco plants

42. a stimulant obtained from coffee beans and tea

Answers to the Practice Test

1. A	2. C	3. B	4. D	5. E
6. D	7. E	8. B	9. A	10. C
11. A	12. D	13. C	14. B	15. D
16. A	17. A	18. D	19. A	20. E
21. D	22. C	23. B	24. C	25. E
26. A	27. A	28. C	29. A	30. B
31. A	32. B	33. D	34. C	35. A
36. B	37. D	38. E	39. C	40. D
41. B	42. A			

Study Goals

- Describe the properties and types of lipids.
- Draw the condensed structural formula of a fatty acid and identify it as saturated or unsaturated.
- Draw the condensed structural formulas of triacylglycerols.
- Draw the condensed structural formula of the product from hydrogenation, hydrolysis, and saponification of triacylglycerols.
- Distinguish between triacylglycerols and glycerophospholipids.
- Describe steroids and their role in bile salts and hormones.
- Describe the lipid bilayer in a cell.

Think About It

1. What are fats used for in the body?

2. What foods are high in fat?

3. What oils are used to produce margarines?

4. What kind of lipid is cholesterol?

Key Terms

Match each of the following key terms with the correct statement shown below:

a. lipid b. fatty acid c. triacylglycerol
d. saponification e. glycerophospholipid f. steroid

1. ____ a lipid consisting of glycerol bonded to two fatty acids and a phosphate group attached to an amino alcohol

2. ____ a type of compound that is not soluble in water, but is soluble in nonpolar solvents

3. ____ the hydrolysis of a triacylglycerol with a strong base producing salts called soaps and glycerol

4. ____ a lipid consisting of glycerol bonded to three fatty acids

5. ____ a lipid containing a multicyclic ring system

6. ____ a long-chain carboxylic acid found in triacylglycerols

Answers   1. e   2. a   3. d   4. c   5. f   6. b

15.1 Lipids

- Lipids are nonpolar compounds that are not soluble in water but are soluble in organic solvents.
- Classes of lipids include waxes, triacylglycerols, glycerophospholipids, and steroids.

(MC) **Tutorial: Classes of Lipids**

♦ **Learning Exercise 15.1**

Match one of the classes of lipids with the composition of lipids below:

  a. wax   **b.** triacylglycerol   **c.** glycerophospholipid   **d.** steroid

1. ____ a fused structure of four cycloalkanes

2. ____ a long chain alcohol and a fatty acid

3. ____ glycerol and three fatty acids

4. ____ glycerol, two fatty acids, phosphate, and choline

Answers   1. d   2. a   3. b   4. c

15.2 Fatty Acids

- Fatty acids are unbranched carboxylic acids that typically contain an even number (12–20) of carbon atoms.
- Fatty acids may be saturated, monounsaturated with one double bond, or polyunsaturated with two or more carbon–carbon double bonds. The double bonds in naturally occurring unsaturated fatty acids are almost always cis.

(MC) **Tutorial: Lipids and Fatty Acids** **Tutorial: Structures and Properties of Fatty Acids** **Self Study Activity: Fats**

♦ **Learning Exercise 15.2A**

Draw the condensed structural formula and skeletal formula of each of the following fatty acids:

1. linoleic acid

2. stearic acid

3. palmitoleic acid

Answers

1. Linoleic acid $CH_3—(CH_2)_4—CH=CH—CH_2—CH=CH—(CH_2)_7—\overset{\overset{\displaystyle O}{\|}}{C}—OH$

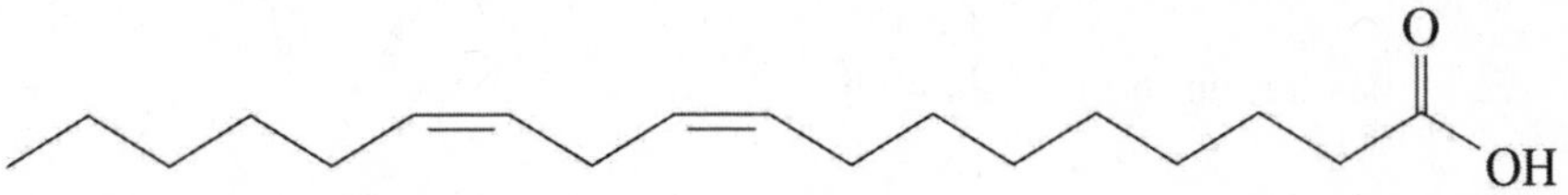

2. Stearic acid $CH_3—(CH_2)_{16}—\overset{\overset{\displaystyle O}{\|}}{C}—OH$

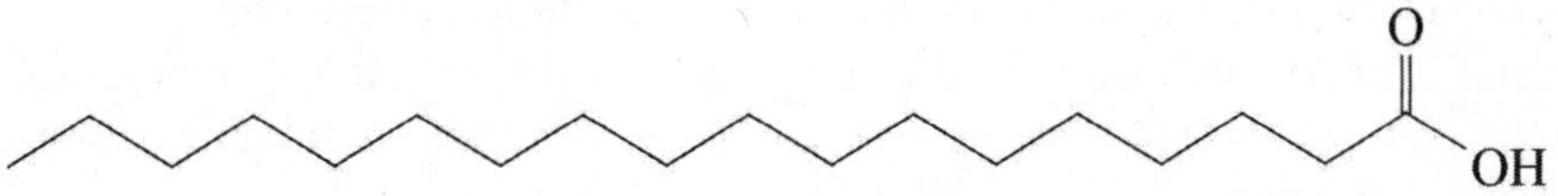

3. Palmitoleic acid $CH_3—(CH_2)_5—CH=CH—(CH_2)_7—\overset{\overset{\displaystyle O}{\|}}{C}—OH$

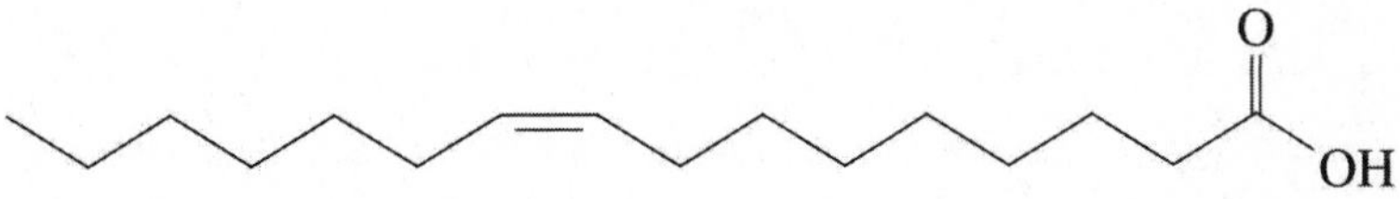

♦ **Learning Exercise 15.2B**

For the fatty acids in Learning Exercise 15.2A, identify which:

1. _____ is the most saturated
2. _____ is the most unsaturated
3. _____ has the lowest melting point
4. _____ has the highest melting point
5. _____ is/are found in vegetables
6. _____ is/are from animal sources

Answers   1. B   2. A   3. A   4. B   5. A, C   6. B

♦ **Learning Exercise 15.2C**

For the following questions, refer to the skeletal formula for oleic acid:

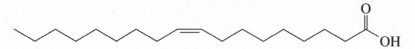

1. Why is the compound an acid?

2. Is it a saturated or unsaturated compound? Why?

3. Is the double bond cis or trans?

4. Is it likely to be a solid or a liquid at room temperature?

5. Why is it not soluble in water?

Answers	1. contains a carboxylic acid group	2. unsaturated; double bond
	3. cis	4. liquid
	5. it has a long hydrocarbon chain	

15.3 Waxes, Fats, and Oils

- A wax is an ester of a long-chain saturated fatty acid and a long-chain alcohol.
- The triacylglycerols in fats and oils are esters of glycerol with three long-chain fatty acids.
- Fats from animal sources contain more saturated fatty acids and have higher melting points than fats found in most vegetable oils.

Self Study Activity: Fats

♦ **Learning Exercise 15.3A**

Draw the condensed structural formula of the wax formed by the reaction of palmitic acid, CH_3—$(CH_2)_{14}$—COOH, and cetyl alcohol, CH_3—$(CH_2)_{14}$—CH_2—OH.

$$
\overset{\displaystyle O}{\overset{\|}{}}
$$

Answer CH_3—$(CH_2)_{14}$—$\overset{\overset{\displaystyle O}{\|}}{C}$—O—$CH_2$—$(CH_2)_{14}$—$CH_3$

♦ **Learning Exercise 15.3B**

Draw the condensed structural formula and write the name of the triacylglycerol formed from the following:

1. glycerol and three palmitic acid molecules, CH_3—$(CH_2)_{14}$—COOH

2. glycerol and three myristic acid molecules, CH_3—$(CH_2)_{12}$—COOH

Answers

1. 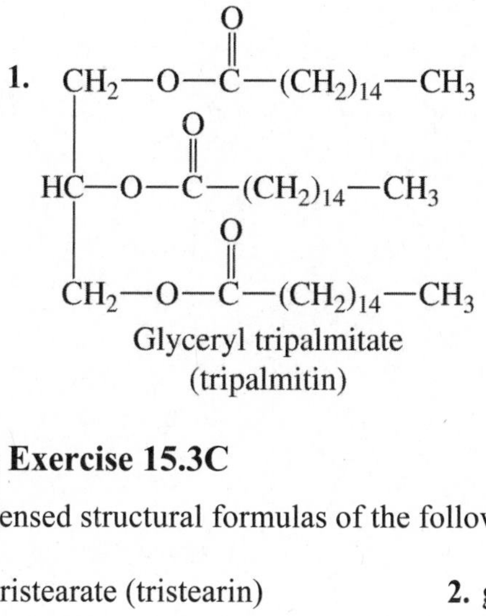
$$
\begin{array}{l}
\quad\quad\quad\quad\quad O \\
\quad\quad\quad\quad\quad \| \\
CH_2-O-C-(CH_2)_{14}-CH_3 \\
\quad\quad\quad\quad O \\
\quad\quad\quad\quad \| \\
HC-O-C-(CH_2)_{14}-CH_3 \\
\quad\quad\quad\quad O \\
\quad\quad\quad\quad \| \\
CH_2-O-C-(CH_2)_{14}-CH_3
\end{array}
$$

Glyceryl tripalmitate
(tripalmitin)

2. 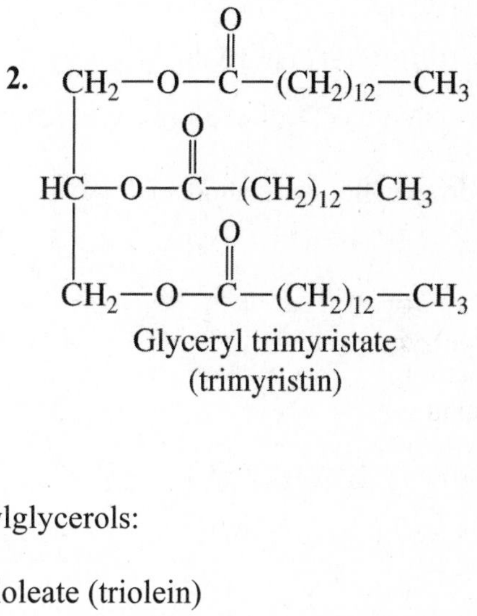
$$
\begin{array}{l}
\quad\quad\quad\quad\quad O \\
\quad\quad\quad\quad\quad \| \\
CH_2-O-C-(CH_2)_{12}-CH_3 \\
\quad\quad\quad\quad O \\
\quad\quad\quad\quad \| \\
HC-O-C-(CH_2)_{12}-CH_3 \\
\quad\quad\quad\quad O \\
\quad\quad\quad\quad \| \\
CH_2-O-C-(CH_2)_{12}-CH_3
\end{array}
$$

Glyceryl trimyristate
(trimyristin)

♦ **Learning Exercise 15.3C**

Draw the condensed structural formulas of the following triacylglycerols:

1. glyceryl tristearate (tristearin)

2. glyceryl trioleate (triolein)

1. 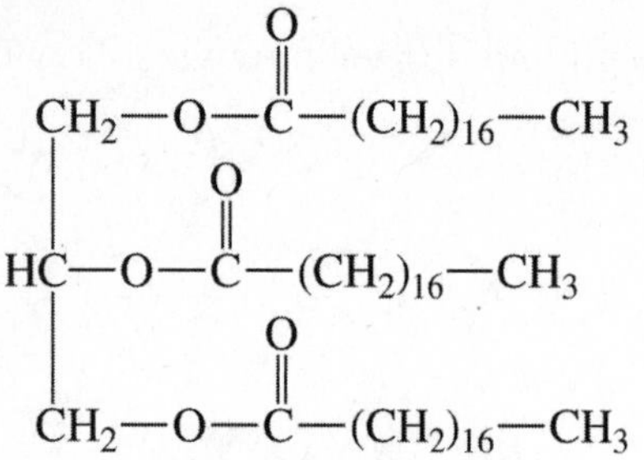
$$CH_2-O-\overset{\displaystyle O}{\overset{\|}{C}}-(CH_2)_{16}-CH_3$$
$$HC-O-\overset{\displaystyle O}{\overset{\|}{C}}-(CH_2)_{16}-CH_3$$
$$CH_2-O-\overset{\displaystyle O}{\overset{\|}{C}}-(CH_2)_{16}-CH_3$$

Glyceryl tristearate
(tristearin)

2. 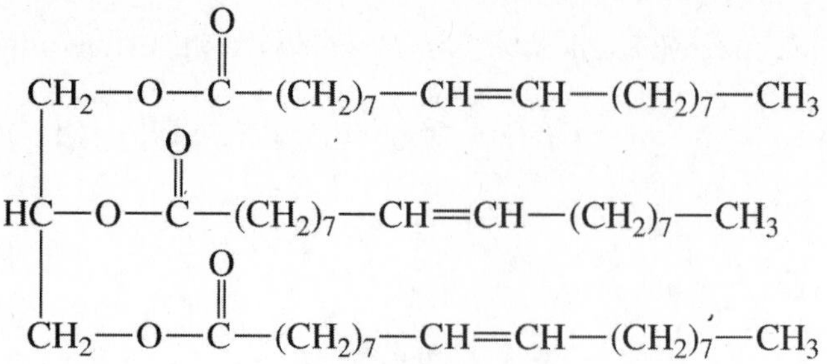
$$CH_2-O-\overset{\displaystyle O}{\overset{\|}{C}}-(CH_2)_7-CH{=}CH-(CH_2)_7-CH_3$$
$$HC-O-\overset{\displaystyle O}{\overset{\|}{C}}-(CH_2)_7-CH{=}CH-(CH_2)_7-CH_3$$
$$CH_2-O-\overset{\displaystyle O}{\overset{\|}{C}}-(CH_2)_7-CH{=}CH-(CH_2)_7-CH_3$$

Glyceryl trioleate
(triolein)

15.4 Chemical Properties of Triacylglycerols

- The hydrogenation of unsaturated fatty acids converts carbon–carbon double bonds to carbon–carbon single bonds.
- The hydrolysis of the ester bonds in fats or oils produces glycerol and fatty acids.
- In saponification, a fat heated with a strong base produces glycerol and the salts of the fatty acids (soaps). The dual polarity of soap permits its solubility in both water and oil.

(MC)

Tutorial: Hydrolysis and Hydrogenation of Triacylglycerols

♦ **Learning Exercise 15.4**

Write the equations for the following reactions of glyceryl trioleate (triolein):

  1. hydrogenation with a nickel catalyst

  2. acid hydrolysis with HCl

  3. saponification with NaOH

Answers

1.

$$CH_2-O-\overset{\overset{\displaystyle O}{\|}}{C}-(CH_2)_7-CH=CH-(CH_2)_7-CH_3$$

$$HC-O-\overset{\overset{\displaystyle O}{\|}}{C}-(CH_2)_7-CH=CH-(CH_2)_7-CH_3 + 3H_2 \xrightarrow{\text{Ni}}$$

$$CH_2-O-\overset{\overset{\displaystyle O}{\|}}{C}-(CH_2)_7-CH=CH-(CH_2)_7-CH_3$$

$$CH_2-O-\overset{\overset{\displaystyle O}{\|}}{C}-(CH_2)_{16}-CH_3$$

$$HC-O-\overset{\overset{\displaystyle O}{\|}}{C}-(CH_2)_{16}-CH_3$$

$$CH_2-O-\overset{\overset{\displaystyle O}{\|}}{C}-(CH_2)_{16}-CH_3$$

2.

$$CH_2-O-\overset{\overset{\displaystyle O}{\|}}{C}-(CH_2)_7-CH=CH-(CH_2)_7-CH_3$$

$$HC-O-\overset{\overset{\displaystyle O}{\|}}{C}-(CH_2)_7-CH=CH-(CH_2)_7-CH_3 + 3H_2O \xrightarrow{\text{H}^+}$$

$$CH_2-O-\overset{\overset{\displaystyle O}{\|}}{C}-(CH_2)_7-CH=CH-(CH_2)_7-CH_3$$

$$CH_2-OH$$
$$HC-OH$$
$$CH_2-OH$$

$$+ 3\ HO-\overset{\overset{\displaystyle O}{\|}}{C}-(CH_2)_7-CH=CH-(CH_2)_7-CH_3$$

3.

$$CH_2-O-\overset{\overset{\displaystyle O}{\|}}{C}-(CH_2)_7-CH=CH-(CH_2)_7-CH_3$$

$$HC-O-\overset{\overset{\displaystyle O}{\|}}{C}-(CH_2)_7-CH=CH-(CH_2)_7-CH_3 + 3NaOH \longrightarrow$$

$$CH_2-O-\overset{\overset{\displaystyle O}{\|}}{C}-(CH_2)_7-CH=CH-(CH_2)_7-CH_3$$

$$CH_2-OH$$
$$HC-OH$$
$$CH_2-OH$$

$$+ 3\ Na^{+\ -}O-\overset{\overset{\displaystyle O}{\|}}{C}-(CH_2)_7-CH=CH-(CH_2)_7-CH_3$$

15.5 Glycerophospholipids

- Glycerophospholipids are esters of glycerol with two fatty acids and a phosphate group attached to an amino alcohol.
- The fatty acids are a nonpolar region, whereas the phosphate group and the ionized amino alcohol make up a polar region.

Tutorial: The Split Personality of Glycerophospholipids

♦ **Learning Exercise 15.5A**

Draw the condensed structural formula of a glycerophospholipid that is formed from two molecules of palmitic acid and serine, an amino alcohol:

$$CH_3—(CH_2)_{14}—\overset{\overset{\displaystyle O}{\|}}{C}OH$$

Palmitic acid

$$HO—CH_2—\overset{\overset{\displaystyle \overset{+}{N}H_3}{|}}{C}H—\overset{\overset{\displaystyle O}{\|}}{C}—O^-$$

Serine

Answer

$$CH_2—O—\overset{\overset{\displaystyle O}{\|}}{C}—(CH_2)_{14}—CH_3$$
$$H\overset{|}{C}—O—\overset{\overset{\displaystyle O}{\|}}{C}—(CH_2)_{14}—CH_3$$
$$CH_2—O—\overset{\overset{\displaystyle \|}{P}}{\underset{\underset{\displaystyle O^-}{|}}{}}—O—CH_2—\overset{\overset{\displaystyle \overset{+}{N}H_3}{|}}{C}H—\overset{\overset{\displaystyle O}{\|}}{C}—O^-$$

♦ **Learning Exercise 15.5B**

Use the following glycerophospholipid to answer parts A to F and 1 to 3:

$$CH_2—O—\overset{\overset{\displaystyle O}{\|}}{C}—(CH_2)_{14}—CH_3$$
$$H\overset{|}{C}—O—\overset{\overset{\displaystyle O}{\|}}{C}—(CH_2)_{14}—CH_3$$
$$CH_2—O—\overset{\overset{\displaystyle O}{\|}}{\underset{\underset{\displaystyle O^-}{|}}{P}}—O—CH_2—CH_2—\overset{+}{N}H_3$$

On the above condensed structural formula, indicate the

    A. two fatty acids   **B.** part from the glycerol molecule

    C. phosphate section   **D.** amino alcohol group

    E. nonpolar region   **F.** polar region

  1. What is the name of the amino alcohol group?   _____

  2. What type of glycerophospholipid is it?   _____

  3. Why is a glycerophospholipid more soluble in water than most lipids? _____

Answers

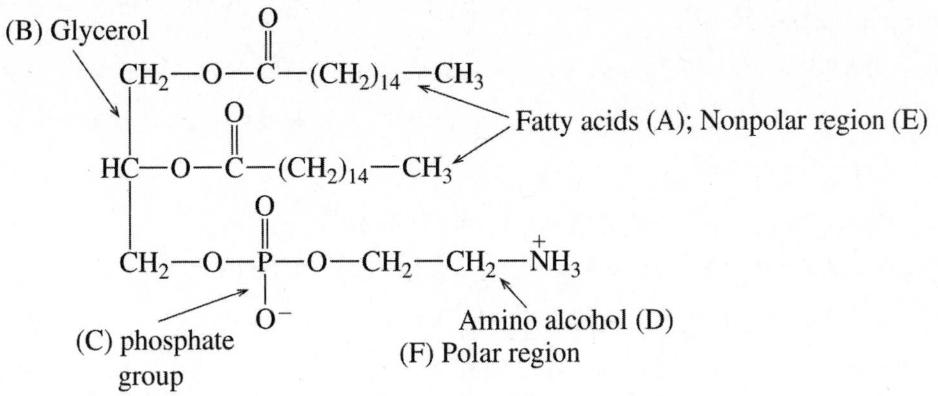

(B) Glycerol

$$CH_2-O-\overset{\overset{\displaystyle O}{\|}}{C}-(CH_2)_{14}-CH_3$$

$$HC-O-\overset{\overset{\displaystyle O}{\|}}{C}-(CH_2)_{14}-CH_3$$

Fatty acids (A); Nonpolar region (E)

$$CH_2-O-\overset{\overset{\displaystyle O}{\|}}{\underset{\underset{\displaystyle O^-}{|}}{P}}-O-CH_2-CH_2-\overset{+}{N}H_3$$

(C) phosphate group

Amino alcohol (D)
(F) Polar region

1. ethanolamine
2. cephalin (ethanolamine glycerophospholipid)
3. The polar portion of the glycerophospholipid is attracted to water, which makes it more soluble in water than other lipids.

15.6 Steroids: Cholesterol, Bile Salts, and Steroid Hormones

- Steroids are lipids containing the steroid nucleus, which is a fused structure of four rings.
- Steroids include cholesterol, bile salts, and steroid hormones.
- The steroid hormones are closely related in structure to cholesterol and depend on cholesterol for their synthesis.
- The sex hormones such as estrogen and testosterone are responsible for sexual characteristics and reproduction.
- The adrenal corticosteroids include aldosterone, which regulates water balance in the kidneys, and cortisone, which regulates glucose levels in the blood.

Tutorial: Cholesterol

♦ **Learning Exercise 15.6A**

1. Draw the structure of the steroid nucleus.

2. Draw the structure of cholesterol.

Answers

1.

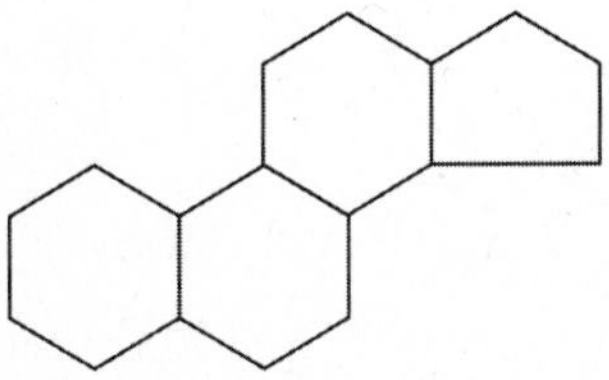

2.

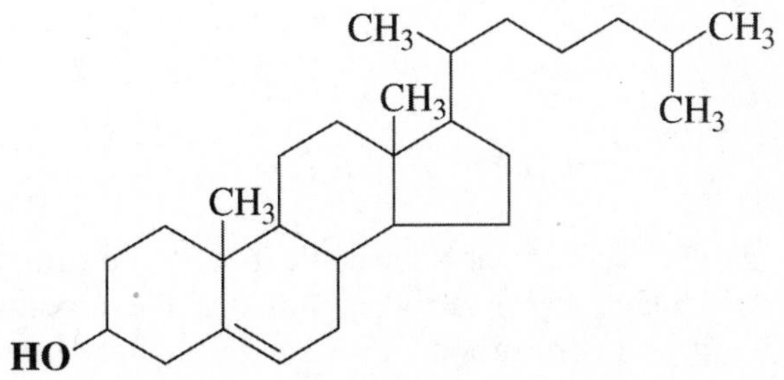

♦ **Learning Exercise 15.6B**

Match each of the following compounds with the statements below:

   a. estrogen   **b.** testosterone   **c.** cortisone   **d.** aldosterone   **e.** bile salts

1. _____ increases the blood level of glucose

2. _____ increases the reabsorption of Na^+ by the kidneys

3. _____ stimulates the development of secondary sex characteristics in females

4. _____ stimulates the retention of water by the kidneys

5. _____ stimulates the secondary sex characteristics in males

6. _____ secreted by the gallbladder into the small intestine to emulsify fats in the diet

Answers   **1.** c   **2.** d   **3.** a   **4.** d   **5.** b   **6.** e

15.7 Cell Membranes

- Cell membranes surround all of our cells and separate the cellular contents from the external aqueous environment.
- A cell membrane is a lipid bilayer composed of two rows of phospholipids such that the nonpolar hydrocarbon tails are in the center and the polar sections are aligned along the outside.
- The center portion of the lipid bilayer consists of nonpolar chains of the fatty acids, with the polar heads at the outer and inner surfaces.
- Molecules of cholesterol, proteins, and carbohydrates are embedded in the lipid bilayer.

♦ **Learning Exercise 15.7**

   a. What is the function of the lipid bilayer in cell membranes?

   b. What type of lipid makes up the lipid bilayer?

   c. What is the general arrangement of the lipids in a lipid bilayer?

Answers

   a. The lipid bilayer separates the contents of a cell from the surrounding aqueous environment.
   b. The lipid bilayer is primarily composed of glycerophospholipids.
   c. The nonpolar hydrocarbon tails are in the center of the bilayer, whereas the polar sections are aligned along the outside of the bilayer.

Checklist for Chapter 15

You are ready to take the Practice Test for Chapter 15. Be sure that you have accomplished the following learning goals for this chapter. If you are not sure, review the section listed at the end of the goal. Then apply your new skills and understanding to the Practice Test.

After studying Chapter 15, I can successfully:

_____ Describe the classes of lipids (15.1).

_____ Draw the condensed structural formula of a fatty acid (15.2).

_____ Identify a fatty acid as saturated or unsaturated (15.2).

_____ Write the condensed structural formula of a wax or triacylglycerol produced by the reaction of a

fatty acid and an alcohol or glycerol (15.3).

_____ Draw the condensed structural formula of the product from the reaction of a triacylglycerol with

hydrogen, an acid, or base (15.4).

_____ Describe the components of glycerophospholipids (15.5).

_____ Describe the structure of a steroid and cholesterol (15.6).

_____ Describe the composition and function of the lipid bilayer in cell membranes (15.7).

Practice Test for Chapter 15

1. An ester of a fatty acid is a type of
   A. carbohydrate   B. lipid   C. protein   D. oxyacid   E. soap

2. A fatty acid that is unsaturated is usually
   A. from animal sources and liquid at room temperature
   B. from animal sources and solid at room temperature
   C. from vegetable sources and liquid at room temperature
   D. from vegetable sources and solid at room temperature
   E. from both vegetable and animal sources and solid at room temperature

3. The following condensed structural formula is a/an:

$$CH_3-(CH_2)_{16}-\overset{\overset{\displaystyle O}{\|}}{C}-OH$$

   A. unsaturated fatty acid   B. saturated fatty acid   C. wax
   D. triacylglycerol   E. steroid

For questions 4 through 7, consider the following compound:

$$CH_2-O-\overset{\overset{\displaystyle O}{\|}}{C}-(CH_2)_{16}-CH_3$$
$$HC-O-\overset{\overset{\displaystyle O}{\|}}{C}-(CH_2)_{16}-CH_3$$
$$CH_2-O-\overset{\overset{\displaystyle O}{\|}}{C}-(CH_2)_{16}-CH_3$$

4. This compound belongs to the family called
   A. wax   B. triacylglycerol   C. glycerophospholipid
   D. fatty acid   E. steroid

5. The molecule shown above was formed by
   A. esterification   B. hydrolysis (acid)   C. saponification
   D. emulsification   E. reduction

6. If this molecule were reacted with a strong base such as NaOH, the products would be
   A. glycerol and fatty acids   B. glycerol and water
   C. glycerol and soap   D. an ester and salts of fatty acids
   E. an ester and fatty acids

7. The compound would be expected to be
   A. saturated and a solid at room temperature
   B. saturated and a liquid at room temperature
   C. unsaturated and a solid at room temperature
   D. unsaturated and a liquid at room temperature
   E. supersaturated and a liquid at room temperature

8. Which are found in glycerophospholipids?
   A. fatty acids   B. glycerol   C. a nitrogen compound
   D. phosphate   E. all of these

For questions 9 and 10, consider the following reaction:

Triacylglycerol + 3NaOH → 3 sodium salts of fatty acids + glycerol

9. The reaction of a triacylglycerol with a strong base such as NaOH is called
   A. esterification   B. lipogenesis   C. hydrolysis
   D. saponification   E. reduction

10. What is another name for the sodium salts of the fatty acids?
    A. margarines   B. fat substitutes   C. soaps   D. perfumes   E. vitamins

For questions 11 through 16, consider the following phosphoglyceride:

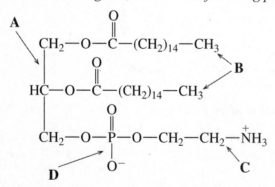

Match the labels with the following:

11. _____ the glycerol portion   12. _____ the phosphate portion   13. _____ the amino alcohol

14. _____ the polar region   15. _____ the nonpolar region

16. This compound belongs in the family called
    A. choline   **B.** cephalin   **C.** fatty acid   **D.** steroid   **E.** triacylglycerol

For questions 17 through 20, classify the following lipids as:

    A. wax   **B.** triacylglycerol   **C.** glycerophospholipid   **D.** steroid   **E.** fatty acid

17. _____ cholesterol

18. _____
$$CH_3-(CH_2)_{14}-\overset{\displaystyle O}{\overset{\|}{C}}-OH$$

19. _____
$$CH_3-(CH_2)_{14}-\overset{\displaystyle O}{\overset{\|}{C}}-O-(CH_2)_{30}-CH_3$$

20. _____ an ester of glycerol with three palmitic acid molecules

For questions 21 through 25, select answers from the following:

    A. testosterone   **B.** estrogen   **C.** prednisone   **D.** cortisone   **E.** aldosterone

21. _____ stimulates the female sexual characteristics

22. _____ increases the retention of water by the kidneys

23. _____ stimulates the male sexual characteristics

24. _____ increases the blood glucose level

25. _____ used medically to reduce inflammation and treat asthma

26. The following compound is a:

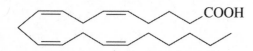

    A. cholesterol   **B.** steroid   **C.** fatty acid   **D.** triacylglycerol   **E.** steroid

27. The lipid bilayer of a cell is composed of
    A. cholesterol   **B.** glycerophospholipids   **C.** proteins   **D.** carbohydrates   **E.** all of these

28. The type of transport that allows chloride ions to move through the integral proteins in the cell membrane is
    A. passive transport   **B.** active transport   **C.** diffusion
    D. facilitated transport   **E.** all of these

29. The movement of small molecules through a cell membrane from a higher concentration to a lower concentration is
    A. passive transport   **B.** active transport   **C.** diffusion
    D. facilitated transport   **E.** A and C

30. The type of lipoprotein that transports cholesterol to the liver for elimination is called

    A. chylomicron   **B.** high-density lipoprotein   **C.** low-density lipoprotein

    D. very-low-density lipoprotein   **E.** all of these

Answers to the Practice Test

1. B	**2.** C	**3.** B	**4.** B	**5.** A
6. C	**7.** A	**8.** E	**9.** D	**10.** C
11. A	**12.** D	**13.** C	**14.** C, D	**15.** B
16. B	**17.** D	**18.** E	**19.** A	**20.** B
21. B	**22.** E	**23.** A	**24.** D	**25.** C
26. C	**27.** E	**28.** D	**29.** E	**30.** B

Amino Acids, Proteins, and Enzymes

Study Goals

- Classify proteins by their functions in the cells.
- Draw the ionized condensed structural formulas of amino acids.
- Draw the ionized condensed structural formulas of di- and tripeptides.
- Identify the structural levels of proteins as primary, secondary, tertiary, and quaternary.
- Describe the effects of denaturation on the structure of proteins.
- Classify enzymes according to the type of reaction they catalyze.
- Describe the lock-and-key and induced-fit models of enzyme action.
- Discuss the effect of changes in temperature, pH, and concentration of substrate on enzyme action.
- Describe the competitive and noncompetitive inhibition of enzymes.
- Identify the types of cofactors that are necessary for enzyme action.

Think About It

1. What are some uses of protein in the body?

2. What are the units that make up a protein?

3. How do you obtain protein in your diet?

4. What are some functions of enzymes in the cells of the body?

5. Why are enzymes sensitive to high temperatures and low or high pH levels?

Key Terms

1. Match the following key terms with the correct statement shown below.

   a. amino acid   **b.** peptide bond   **c.** denaturation
   d. primary structure   **e.** zwitterion

1. _____ the order of amino acids in a protein

2. _____ the ionized structure of an amino acid that has a net charge of zero

3. _____ the bond that connects amino acids in peptides and proteins

4. _____ the loss of secondary and tertiary protein structure caused by agents such as heat and acid

5. _____ the building block of proteins

Answers   1. d   2. e   3. b   4. c   5. a

2. Match the following key terms with the correct statement shown below.

   a. lock-and-key theory   **b.** vitamin   **c.** inhibitor   **d.** enzyme   **e.** active site

1. _____ the portion of an enzyme structure where a substrate undergoes reaction

2. _____ a protein that catalyzes a biological reaction in the cells

3. _____ a model of enzyme action in which the substrate exactly fits the shape of an enzyme like a key fits into a lock

4. _____ a substance that makes an enzyme inactive by interfering with its ability to react with a substrate

5. _____ an organic compound essential for normal health and growth that must be obtained from the diet

Answers   1. e   2. d   3. a   4. c   5. b

16.1 Proteins and Amino Acids

- Some proteins are enzymes or hormones, while others are important in structure, transport, protection, storage, and contraction of muscles.
- A group of 20 amino acids provides the molecular building blocks of proteins.
- In an ionized amino acid, a central (alpha) carbon is usually attached to an ammonium group, $-NH_3^+$, a carboxylate group ($-COO^-$), a hydrogen atom ($-H$), and a side chain or R group, which is unique for each amino acid.
- Each specific R group determines if an amino acid is nonpolar, polar (neutral), acidic, or basic. Nonpolar amino acids contain hydrocarbon side chains, whereas polar amino acids contain electro-negative atoms such as oxygen ($-OH$) or sulfur ($-SH$). Acidic side chains contain a carboxylate group ($-COO^-$), and basic side chains contain an ammonium group ($-NH_3^+$).

(MC)

Tutorial: Protein Building Blocks
Tutorial: Proteins "R" Us
Self Study Activity: Functions of Proteins

♦ **Learning Exercise 16.1A**

Match one of the following functions of a protein with the examples below:

   a. structural   **b.** contractile   **c.** storage   **d.** transport

   e. hormonal   **f.** enzyme   **g.** protection

1. _____ hemoglobin, carries oxygen in blood   2. _____ collagen, makes up connective tissue

3. _____ amylase, hydrolyzes starch   4. _____ immunoglobulin, helps fight infections

5. _____ egg albumin, a protein in egg white   6. _____ keratin, a major protein of hair

7. _____ vasopressin, regulates blood pressure   8. _____ lipoprotein, carries lipids in blood

Answers   **1.** d   **2.** a   **3.** f   **4.** g
               5. c   **6.** a   **7.** e   **8.** d

♦ Learning Exercise 16.1B

Using the appropriate R group, complete the condensed structural formula of each of the following amino acids. Indicate whether the amino acid would be nonpolar, polar (neutral), acidic, or basic.

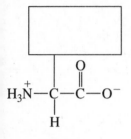

Glycine

_____

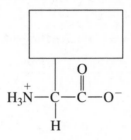

Alanine

_____

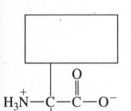

Serine

_____

Aspartic acid

_____

Answers

$$H_3\overset{+}{N}-\underset{\underset{H}{|}}{\overset{\overset{H}{|}}{C}}-\overset{\overset{O}{\|}}{C}-O^-$$

Nonpolar

$$H_3\overset{+}{N}-\underset{\underset{H}{|}}{\overset{\overset{CH_3}{|}}{C}}-\overset{\overset{O}{\|}}{C}-O^-$$

Nonpolar

$$CH_2-OH$$
$$H_3\overset{+}{N}-\underset{\underset{H}{|}}{\overset{|}{C}}-\overset{\overset{O}{\|}}{C}-O^-$$

Polar (neutral)

$$CH_2-\overset{\overset{O}{\|}}{C}-O^-$$
$$H_3\overset{+}{N}-\underset{\underset{H}{|}}{\overset{|}{C}}-\overset{\overset{O}{\|}}{C}-O^-$$

Acidic

16.2 Amino Acids as Acids and Bases

- Amino acids exist as dipolar ions called zwitterions, which are neutral at the isoelectric point (pI).
- A zwitterion has a positive charge at pH levels below its pI and a negative charge at pH levels above its pI.

Tutorial: pH, pI, and Amino Acid Ionization

Study Note

Example: Glycine has an isoelectric point at a pH of 6.0. Write the condensed structural formula of glycine at its isoelectric point (pI), and at pH levels above and below its isoelectric point.

Solution: In more acidic solutions, glycine has a net positive charge, and in more basic solutions, a net negative charge.

$$\overset{+}{H_3N}-CH_2-COOH \xleftarrow{\;H^+\;} \overset{+}{H_3N}-CH_2-COO^- \xrightarrow{\;OH^-\;} H_2N-CH_2-COO^-$$

below pI	*at pI*	*above pI*

◆ Learning Exercise 16.2

Draw the ionized structural formulas of the amino acids under each of the given conditions:

Zwitterion	low pH	high pH
Alanine		
Serine		

Answers

Zwitterion	low pH	high pH
Alanine $\overset{CH_3}{\underset{\overset{+}{H_3N}-CH-COO^-}{\mid}}$	$\overset{CH_3}{\underset{\overset{+}{H_3N}-CH-COOH}{\mid}}$	$\overset{CH_3}{\underset{H_2N-CH-COO^-}{\mid}}$
Serine $\overset{CH_2-OH}{\underset{\overset{+}{H_3N}-CH-COO^-}{\mid}}$	$\overset{CH_2OH}{\underset{\overset{+}{H_3N}-CH-COOH}{\mid}}$	$\overset{CH_2OH}{\underset{H_2N-CH-COO^-}{\mid}}$

16.3 Formation of Peptides

- A peptide bond is an amide bond between the carboxylate group (COO⁻) of one amino acid and the ammonium, NH_3^+, group of the second.

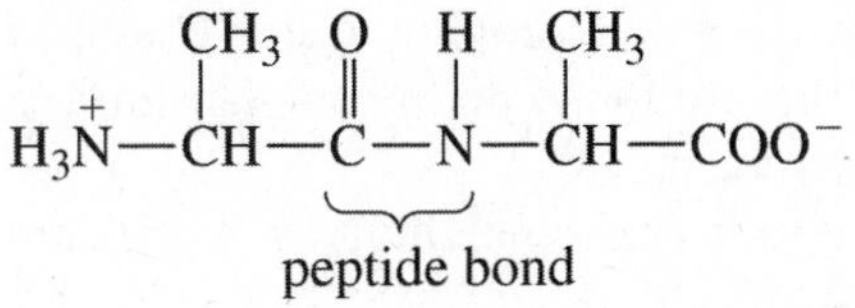

$$\underset{}{H_3\overset{+}{N}}-CH-\overset{\displaystyle O}{\overset{\displaystyle \|}{C}}-N-CH-COO^-$$

CH₃   O   H   CH₃

peptide bond

- Short chains of amino acids are called peptides. Long chains of amino acids are called proteins.

Self Study Activity: Structure of Proteins
Tutorial: Peptides Are Chains of Amino Acids
Tutorial: Peptide Bonds: Acid Meets Amino

♦ Learning Exercise 16.3

Draw the ionized condensed structural formulas of the following di- and tripeptides:

1. serylglycine

2. cystylvaline

3. Gly-Ser-Cys

Answers

1. 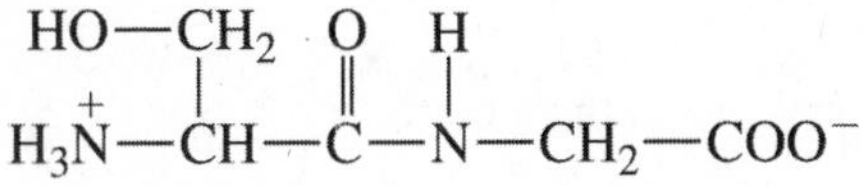
$$HO-CH_2 \quad O \quad H$$
$$H_3\overset{+}{N}-CH-\overset{O}{\overset{\|}{C}}-N-CH_2-COO^-$$

2.
$$HS-CH_2 \quad O \quad H \quad \overset{CH_3}{\underset{}{CH}}-CH_3$$
$$H_3\overset{+}{N}-CH-\overset{O}{\overset{\|}{C}}-N-CH-COO^-$$

3.
$$\quad O \quad HO-CH_2 \quad O \quad CH_2-SH$$
$$H_3\overset{+}{N}-CH_2-\overset{O}{\overset{\|}{C}}-N-CH-\overset{O}{\overset{\|}{C}}-N-CH-COO^-$$
$$\qquad\qquad\quad H \qquad\qquad\quad H$$

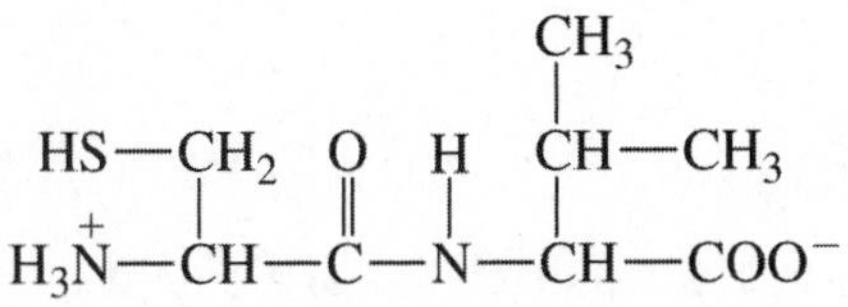

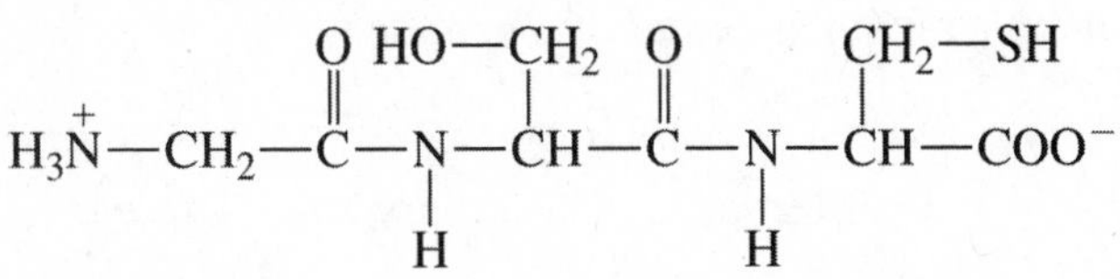

16.4 Levels of Protein Structure

- The primary structure of a protein is the sequence of amino acids.
- In the secondary structure, hydrogen bonds between different sections of the peptide produce a characteristic shape such as an α helix, β-pleated sheet, or triple helix.
- In a tertiary structure, amino acids with hydrophobic side chains are found on the inside and amino acids with hydrophilic side chains are found on the outside surface. The tertiary structure is stabilized by interactions between the side chains.
- In globular proteins, the polypeptide chain including its α-helical and β-pleated sheet regions folds onto itself to form a tertiary structure.
- In a quaternary structure, two or more subunits must combine for biological activity. They are held together by the same interactions found in tertiary structures.
- Denaturation of a protein occurs when there is a change that destroys the secondary and tertiary structure (but not the primary structure) of the protein until biological activity is lost.

(MC)

Self Study Activity: Structure of Proteins
Self Study Activity: Primary and Secondary Structure
Tutorial: The Shapes of Protein Chains: Helices and Sheets
Self Study Activity: Tertiary and Quaternary Structure
Tutorial: Levels of Structure in Proteins
Tutorial: Protein Demolition
Tutorial: Understanding Protein Degradation

◆ **Learning Exercise 16.4A**

Identify the following descriptions of protein structure as primary or secondary:

1. _____ hydrogen bonding forming an alpha (α) helix

2. _____ hydrogen bonding occuring between C=O and N—H within a peptide chain

3. _____ the order of amino acids in a peptide

4. _____ hydrogen bonds between protein chains forming a pleated-sheet structure

Answers   **1.** secondary   **2.** secondary   **3.** primary   **4.** secondary

◆ **Learning Exercise 16.4B**

Identify the following descriptions of protein structure as tertiary or quaternary:

1. _____ a disulfide bond joining distant parts of a peptide

2. _____ the combination of four protein subunits

3. _____ hydrophilic side groups seeking contact with water

4. _____ a salt bridge forming between two oppositely charged side chains of a peptide

5. _____ hydrophobic side groups forming a nonpolar center of a peptide

Answers   **1.** tertiary   **2.** quaternary   **3.** tertiary   **4.** tertiary   **5.** tertiary

◆ **Learning Exercise 16.4C**

Indicate the denaturing agent in the following examples:

    a. heat   **b.** pH change   **c.** organic solvent

    d. heavy metal ions   **e.** agitation

1. _____ placing surgical instruments in a 120 °C autoclave

2. _____ whipping cream to make a dessert topping

3. _____ applying tannic acid to a burn

4. _____ placing $AgNO_3$ drops in the eyes of newborns

5. _____ using alcohol to disinfect a wound

6. _____ using *lactobacillus* bacteria culture to produce acid that converts milk to yogurt

Answers   **1.** a   **2.** e   **3.** b   **4.** d   **5.** c   **6.** b

16.5 Enzymes

- Enzymes are globular proteins that act as biological catalysts.
- Enzymes accelerate the rate of biological reactions by lowering the activation energy of a reaction.
- The names of most enzymes are indicated by their *ase* endings.
- Enzymes are classified by the type of reaction they catalyze: oxidoreductase, hydrolase, isomerase, transferase, lyase, or ligase.

Tutorial: Enzymes and Activation Energy

◆ **Learning Exercise 16.5A**

Indicate whether each of the following characteristics of an enzyme is *true* or *false:*

An enzyme

1. _____ is a biological catalyst

2. _____ does not change the equilibrium position of a reaction

3. _____ is obtained from the diet

4. _____ greatly increases the rate of a cellular reaction

5. _____ is needed for every reaction that takes place in the cell

6. _____ catalyzes at a faster rate at higher temperatures

7. _____ lowers the activation energy of a biological reaction

8. _____ increases the rate of the forward reaction, but not the reverse

Answers   **1.** T   **2.** T   **3.** F   **4.** T

                  5. T   **6.** F   **7.** T   **8.** F

♦ **Learning Exercise 16.5B**

Match the name of each of the following enzymes with the description of the reaction:

1. dehydrogenase   2. oxidase   3. peptidase

4. decarboxylase   5. esterase   6. transaminase

a. _____ hydrolyzes the ester bonds in triacylglycerols to yield fatty acids and glycerol

b. _____ removes hydrogen from a substrate

c. _____ removes CO_2 from a substrate

d. _____ decomposes hydrogen peroxide to water and oxygen

e. _____ hydrolyzes peptide bonds during the digestion of proteins

f. _____ transfers an amino (NH_2) group from an amino acid to an α-keto acid

Answers   **a.** 5   **b.** 1   **c.** 4   **d.** 2   **e.** 3   **f.** 6

♦ **Learning Exercise 16.5C**

Match the classification for enzymes with each of the following types of reactions:

1. oxidoreductase   2. transferase   3. hydrolase

4. lyase   5. isomerase   6. ligase

a. _____ combines small molecules using energy from ATP

b. _____ transfers phosphate groups

c. _____ hydrolyzes a disaccharide into two glucose units

d. _____ converts a substrate to an isomer

e. _____ adds hydrogen to a substrate

f. _____ removes H_2O from a substrate

g. _____ adds oxygen to a substrate

h. _____ converts a cis structure to a trans structure

Answers   **a.** 6   **b.** 2   **c.** 3   **d.** 5
             e. 1   **f.** 4   **g.** 1   **h.** 5

16.6 Enzyme Action

- Within the structure of the enzyme, there is a small pocket called the active site, which has a specific shape that fits a specific substrate.
- In the lock-and-key model or the induced-fit model, an enzyme and substrate form an enzyme–substrate complex so the reaction of the substrate can be catalyzed at the active site.

(MC)

Tutorial: Enzymes and the Lock-and-Key Model

♦ **Learning Exercise 16.6A**

Match the terms (A) active site, (B) substrate, (C) enzyme–substrate complex, (D) lock-and-key, and (E) induced-fit with the following descriptions:

1. _____ the combination of an enzyme with a substrate

2. _____ a model of enzyme action in which the rigid shape of the active site exactly fits the shape of the substrate

3. _____ has a tertiary structure that fits the structure of the active site

4. _____ a model of enzyme action in which the shape of the active site adjusts to fit the shape of a substrate

5. _____ the portion of an enzyme that binds to the substrate and catalyzes the reaction

Answers   **1.** C   **2.** D   **3.** B   **4.** E   **5.** A

♦ **Learning Exercise 16.6B**

Write an equation to illustrate the following:

1. The formation of an enzyme–substrate complex.

2. The conversion of enzyme–substrate complex to product.

Answers   **1.** $E + S \rightleftarrows ES$ **2.** $ES \rightarrow E + P$

16.7 Factors Affecting Enzyme Activity

- Enzymes are most effective at optimum temperature and pH. The rate of an enzyme reaction decreases considerably at temperatures and pH above or below the optimum.
- An enzyme can be made inactive by changes in pH, temperature, or by chemical compounds called inhibitors.
- An increase in substrate concentration increases the reaction rate of an enzyme-catalyzed reaction until all the enzyme molecules combine with substrate.
- A competitive inhibitor has a structure similar to the substrate and competes for the active site. When the active site is occupied by a competitive inhibitor, the enzyme cannot catalyze the reaction of the substrate.
- A noncompetitive inhibitor attaches elsewhere on the enzyme, changing the shape of both the enzyme and the active site. As long as the noncompetitive inhibitor is attached to the enzyme, the altered active site cannot bind with the substrate.

Tutorial: Enzyme and Substrate Concentrations
Tutorial: Enzyme Inhibition

♦ Learning Exercise 16.7A

An enzyme has an optimum pH of 7 and and optimum temperature of 37 °C.

Draw a graph to represent the effects of each of the following on the reaction rate of the enzyme. Indicate the optimum pH and optimum temperature.

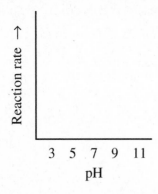
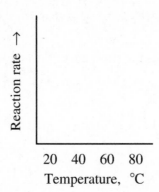

Indicate the effect of each of the following on the reaction rate of the reaction:

    a. increases   **b.** decreases   **c.** not changed

1. _____ adding more substrate when an excess of enzyme is present

2. _____ running the reaction at pH 8

3. _____ lowering the temperature to 0 °C

4. _____ running the reaction at 85 °C

5. _____ increasing the concentration of substrate when all the enzyme is bonded to substrate

6. _____ adjusting pH to the optimum

Answers

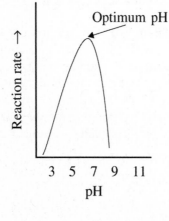
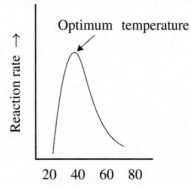

    1. a   **2.** b   **3.** b   **4.** b   **5.** c   **6.** a

♦ **Learning Exercise 16.7B**

Identify each of the following as characteristic of competitive inhibition (C) or noncompetitive inhibition (N):

1. _____ An inhibitor binds to the surface of the enzyme away from the active site.

2. _____ An inhibitor resembling the substrate molecule blocks the active site on the enzyme.

3. _____ The action of this inhibitor can be reversed by adding more substrate.

4. _____ Increasing substrate concentration does not change the effect of this inhibitor.

5. _____ Sulfanilamide stops bacterial infections because its structure is similar to PABA (p-aminobenzoic acid), which is essential for bacterial growth.

Answers   **1.** N   **2.** C   **3.** C   **4.** N   **5.** C

16.8 Enzyme Cofactors

- Simple enzymes are biologically active as a protein only, whereas other enzymes require a cofactor.
- A cofactor may be a metal ion such as Cu^{2+} or Fe^{2+}, or an organic compound called a coenzyme, usually a vitamin.
- Vitamins are organic molecules that are essential for proper health.
- Vitamins must be obtained from the diet because they are not synthesized in the body.
- Vitamins B and C are water-soluble; vitamins A, D, E, and K are fat-soluble.
- Many water-soluble vitamins function as coenzymes.

(MC)

Tutorial: Enzyme Cofactors and Vitamins

♦ **Learning Exercise 16.8A**

Indicate whether each statement describes a simple enzyme or a protein that requires a cofactor.

1. _____ an enzyme consisting only of protein

2. _____ an enzyme requiring magnesium ions for activity

3. _____ an enzyme containing a sugar group

4. _____ an enzyme that gives only amino acids upon hydrolysis

5. _____ an enzyme that requires zinc ions for activity

Answers   1. simple   2. requires a cofactor   3. requires a cofactor
             4. simple   5. requires a cofactor

♦ **Learning Exercise 16.8B**

Identify the water-soluble vitamin associated with each of the following:

a. pantothenic acid (B_5)   **b.** riboflavin (B_2)   **c.** niacin (B_3)   **d.** ascorbic acid (C)

1. _____ collagen formation   2. _____ FAD and FMN

3. _____ NAD^+ 4. _____ Coenzyme A

Answers   **1.** d   **2.** b   **3.** c   **4.** a

♦ **Learning Exercise 16.8C**

Identify the fat-soluble vitamin associated with each of the following:

    a. vitamin A   **b.** vitamin D   **c.** vitamin E   **d.** vitamin K

    1. _____ blood clotting   **2.** _____ vision

    3. _____ antioxidant   **4.** _____ absorption of calcium

Answers   **1.** d   **2.** a   **3.** c   **4.** b

Checklist for Chapter 16

You are ready to take the Practice Test for Chapter 16. Be sure that you have accomplished the following learning goals for this chapter. If you are not sure, review the section listed at the end of the goal. Then apply your new skills and understanding to the Practice Test.

After studying Chapter 16, I can successfully:

_____ Classify proteins by their functions in the cells (16.1).

_____ Draw the ionized condensed structural formula for an amino acid (16.1).

_____ Draw the ionized condensed structural formula of an amino acid at its pI and at pH values above or below the pI (16.2).

_____ Describe a peptide bond; draw the structure for a peptide (16.3).

_____ Distinguish between the primary and secondary structures of a protein (16.4).

_____ Distinguish between the tertiary and quaternary structures of a protein (16.4).

_____ Describe the ways that denaturation affects the structure of a protein (16.4).

_____ Classify enzymes according to the type of reaction they catalyze (16.5).

_____ Describe the lock-and-key and induced-fit models of enzyme action (16.6).

_____ Discuss the effect of changes in temperature, pH, and concentration of substrate on enzyme action (16.7).

_____ Describe the competitive and noncompetitive inhibition of enzymes (16.7).

_____ Identify the types of cofactors that are necessary for enzyme action (16.8).

Practice Test for Chapter 16

1. Which amino acid is nonpolar?
    A. serine   **B.** aspartic acid   **C.** valine   **D.** cysteine   **E.** lysine

2. Which amino acid will form disulfide cross-links in a tertiary structure?
    A. serine   **B.** aspartic acid   **C.** valine   **D.** cysteine   **E.** lysine

3. Which amino acid has a basic side chain?
    A. serine   **B.** aspartic acid   **C.** valine   **D.** cysteine   **E.** lysine

4. All amino acids
    A. have the same side chains   **B.** form zwitterions
    C. have the same isoelectric points   **D.** show hydrophobic tendencies
    E. are essential amino acids

5. Essential amino acids
    A. are the amino acids that must be supplied by the diet
    B. are not synthesized by the body
    C. are missing in incomplete proteins
    D. are present in proteins from animal sources
    E. all of these

For questions 6 through 9, use the following to answer questions for the amino acid alanine.

          CH_3 CH_3 CH_3
                   | | |
A. $\overset{+}{H_3N}-CH-COO^-$ **B.** $H_2N-CH-COO^-$ **C.** $\overset{+}{H_3N}-CH-COOH$

6. _____ alanine in its zwitterion form

7. _____ alanine at a low pH

8. _____ alanine at a high pH

9. _____ alanine at its isoelectric point

10. The sequence Tyr-Ala-Gly
    A. is a tripeptide   **B.** has two peptide bonds   **C.** tyrosine is the N terminal amino acid
    D. glycine is the C terminal amino acid   **E.** All of these

11. The type of bonding expected between lysine and aspartic acid is a/an
    A. salt bridge   **B.** hydrogen bond   **C.** disulfide bond
    D. hydrophobic interaction   **E.** hydrophilic interaction

12. What type of bond is used to form the α helix structure of a protein?
    A. peptide bond   **B.** hydrogen bond   **C.** ionic bond
    D. disulfide bond   **E.** hydrophobic interaction

13. What type of bonding places portions of the protein chain in the center of a tertiary structure?
    A. peptide bonds   **B.** ionic bonds   **C.** disulfide bonds
    D. hydrophobic interactions   **E.** hydrophilic interactions

For questions 14 through 18, identify the protein structural levels that each of the following statements describe:

    A. primary   **B.** secondary   **C.** tertiary   **D.** quaternary   **E.** pentenary

14. _____ peptide bonds   **15.** _____ a pleated sheet

16. _____ two or more protein subunits   **17.** _____ an α helix

18. _____ disulfide bonds

For questions 19 through 22, match the function of a protein with each example.

19. _____ enzyme

    A. myoglobin in the muscles
    B. α-keratin in skin

20. _____ structural

    C. peptidase for protein hydrolysis
    D. casein in milk

21. _____ transport

22. _____ storage

23. Denaturation of a protein
    A. occurs at a pH of 7   **B.** causes a change in protein structure
    C. hydrolyzes a protein   **D.** oxidizes the protein
    E. adds amino acids to a protein

24. Which of the following will <u>not</u> cause denaturation?
    A. $0\,°C$ **B.** $AgNO_3$ **C.** $80\,°C$ **D.** ethanol   **E.** pH 1

25. Enzymes are
    A. biological catalysts   **B.** polysaccharides   **C.** insoluble in water
    D. always contain a cofactor   **E.** named with an "ose" ending

Classify the enzymes described in questions 26 through 29 as (A) simple or (B) requiring a cofactor.

26. _____ an enzyme that yields amino acids and a glucose molecule on analysis

27. _____ an enzyme consisting of protein only

28. _____ an enzyme requiring zinc ion for activation

29. _____ an enzyme containing vitamin K

For questions 30 through 34, select answers from the following: (E = enzyme; S = substrate; P = product).
    A. $S \rightarrow P$ **B.** $ES \rightarrow E + S$ **C.** $E + S \rightleftarrows ES$
    D. $ES \rightarrow EP$ **E.** $EP \rightarrow ES$

30. _____ the enzymatic reaction occurring at the active site

31. _____ the release of product from the enzyme

32. _____ the first step in the lock-and-key theory of enzyme action

33. _____ the formation of the enzyme–substrate complex

34. _____ the final step in the lock-and-key theory of enzyme action

For questions 35 through 39, match the names of enzymes with a reaction they each catalyze.
    A. decarboxylase   **B.** isomerase   **C.** dehydrogenase
    D. lipase   **E.** sucrase

35. _____ $CH_3-\underset{\underset{\textstyle OH}{|}}{CH}-COOH \longrightarrow CH_3-\underset{\underset{\textstyle O}{\|}}{C}-COOH$

36. _____ sucrose + H_2O → glucose and fructose

37. _____ $CH_3-\overset{\overset{\displaystyle O}{\|}}{C}-COOH \rightarrow CH_3-COOH + CO_2$

38. _____ fructose → glucose

39. _____ triglyceride + 3 H_2O → fatty acids and glycerol

For questions 40 through 44, select your answers from A, B, or C.

   A. increases the rate of reaction   **B.** decreases the rate of reaction
   C. denatures the enzyme and no reaction occurs

40. _____ setting the reaction tube in a beaker of water at 100 °C

41. _____ adding more substrate to the reaction vessel

42. _____ running the reaction at 10 °C

43. _____ adding ethanol to the reaction system

44. _____ adjusting the pH to optimum pH

For questions 45 through 47, identify each description of inhibition as

   A. competitive   **B.** noncompetitive

45. _____ an alteration in the conformation of the enzyme

46. _____ a molecule closely resembling the substrate interferes with activity

47. _____ the inhibition can be reversed by increasing substrate concentration

48. _____ the presence of zinc in the enzyme alcohol dehydrogenase classifies the enzyme as
   A. simple   **B.** requiring a cofactor   **C.** hormonal
   D. structural   **E.** secondary

Answers to the Practice Test

1. C	**2.** D	**3.** E	**4.** B	**5.** E
6. A	**7.** C	**8.** B	**9.** A	**10.** E
11. A	**12.** B	**13.** D	**14.** A	**15.** B
16. D	**17.** B	**18.** C, D	**19.** C	**20.** B
21. A	**22.** D	**23.** B	**24.** A	**25.** A
26. B	**27.** A	**28.** B	**29.** B	**30.** D
31. B	**32.** C	**33.** C	**34.** B	**35.** C
36. E	**37.** A	**38.** B	**39.** D	**40.** C
41. A	**42.** B	**43.** C	**44.** A	**45.** B
46. A	**47.** A	**48.** B		

Nucleic Acids and Protein Synthesis

Study Goals

- Draw the structures of the bases, sugars, and nucleotides in DNA and RNA.
- Describe the structures of DNA and RNA.
- Explain the process of DNA replication.
- Describe the transcription process during the synthesis of mRNA.
- Use the codons in the genetic code to describe protein synthesis.
- Explain how an alteration in the DNA sequence can lead to mutations in proteins.
- Explain how retroviruses use reverse transcription to synthesize DNA.

Think About It

1. Where is DNA in your cells?

2. How does DNA determine your height, or the color of your hair or eyes?

3. What is the genetic code?

4. How does a mutation occur?

Key Terms

Match the statements shown below with the following key terms.

    a. DNA   **b.** RNA   **c.** double helix   **d.** mutation   **e.** transcription

1. _____ the formation of mRNA to carry genetic information from DNA to protein synthesis

2. _____ the genetic material containing nucleotides and bases adenine, cytosine, guanine, and thymine

3. _____ the shape of DNA with a sugar–phosphate backbone and base pairs linked in the center

4. _____ a change in the DNA base sequence that may alter the shape and function of a protein

5. _____ a type of nucleic acid with a single strand of nucleotides of adenine, cytosine, guanine, and uracil

17.1 Components of Nucleic Acids

- Nucleic acids are composed of bases, five-carbon sugars, and phosphate groups.
- In DNA, the bases are adenine, thymine, guanine, or cytosine. In RNA, uracil replaces thymine.
- In DNA, the sugar is deoxyribose; in RNA the sugar is ribose.
- A nucleoside is composed of a base and a sugar.
- A nucleotide is composed of three parts: a base, a sugar, and a phosphate group.

Tutorial: Nucleic Acid Building Blocks

♦ **Learning Exercise 17.1A**

1. Write the names and abbreviations for the bases in each of the following:

DNA _____

RNA _____

2. Write the name of the sugar in each of the following nucleotides:

DNA _____

RNA _____

Answers   1. DNA: adenine (A), thymine (T), guanine (G), cytosine (C)
                     RNA: adenine (A), uracil (U), guanine (G), cytosine (C)
                 2. DNA: deoxyribose, RNA: ribose

♦ **Learning Exercise 17.1B**

Name each of the following and classify it as a purine or a pyrimidine:

a.

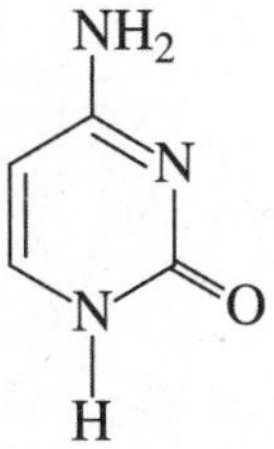

b.

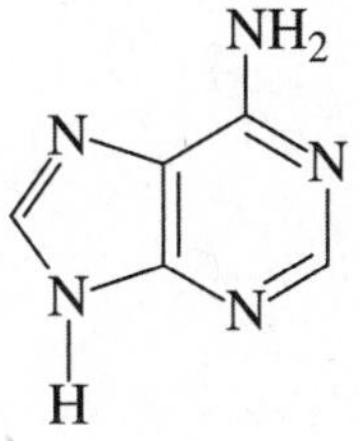

c.

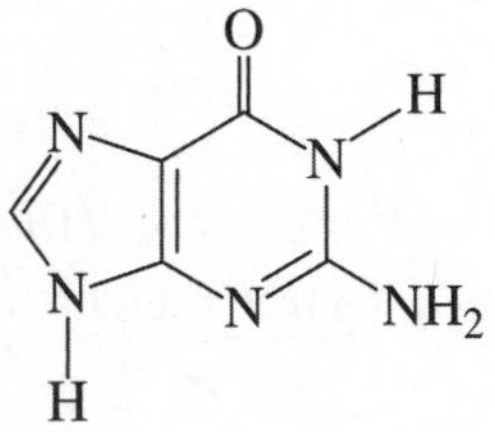

d.

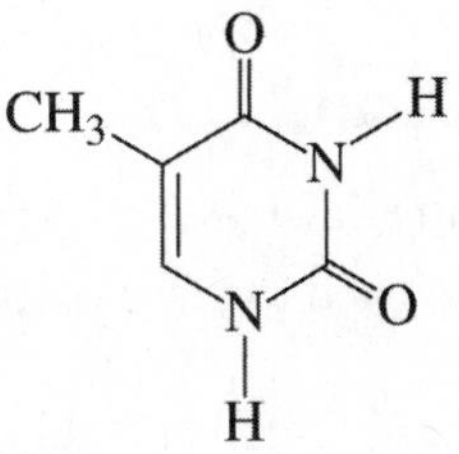

Answers   **a.** cytosine; pyrimidine   **b.** adenine; purine
            c. guanine; purine   **d.** thymine; pyrimidine

♦ **Learning Exercise 17.1C**

Identify the nucleic acid (DNA or RNA) in which each of the following are found:

1. _____ adenosine-5′-monophosphate   2. _____ guanosine-5′-monophosphate

3. _____ dCMP   4. _____ cytidine-5′-monophosphate

5. _____ deoxythymidine-5′-monophosphate   6. _____ UMP

7. _____ dGMP   8. _____ deoxyadenosine-5′-monophosphate

Answers   **1.** RNA   **2.** RNA   **3.** DNA   **4.** RNA
            5. DNA   **6.** RNA   **7.** DNA   **8.** DNA

♦ **Learning Exercise 17.1D**

Write the structural formula for deoxyadenosine-5′-monophosphate. Indicate the 5′- and the 3′-carbon atoms on the sugar.

Answer

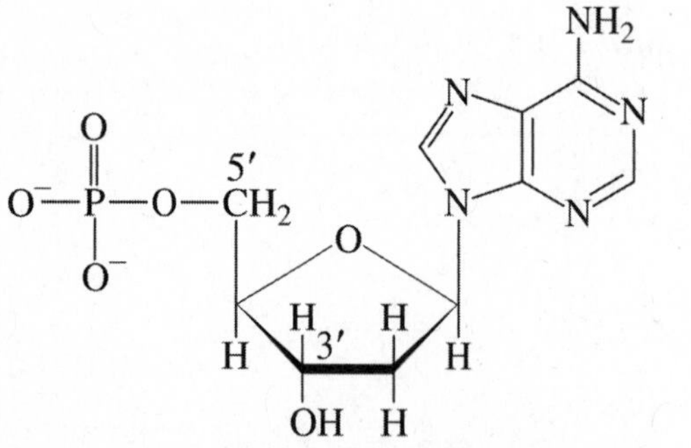

Deoxyadenosine-5′-monophosphate (dAMP)

17.2 Primary Structure of Nucleic Acids

- Nucleic acids are polymers of nucleotides in which the —OH group on the 3′-carbon of a sugar in one nucleotide bonds to the phosphate group attached to the 5′-carbon of a sugar in the adjacent nucleotide.

Self Study Activity: DNA and RNA Structure

♦ **Learning Exercise 17.2A**

Write the structure of a dinucleotide that consists of cytosine-5′-monophosphate (free 5′-phosphate) bonded to guanosine-5′-monophosphate (free 3′-hydroxyl). Identify each nucleotide, the phosphodiester bond, the free 5′-phosphate group, and the free 3′-hydroxyl group.

Answer

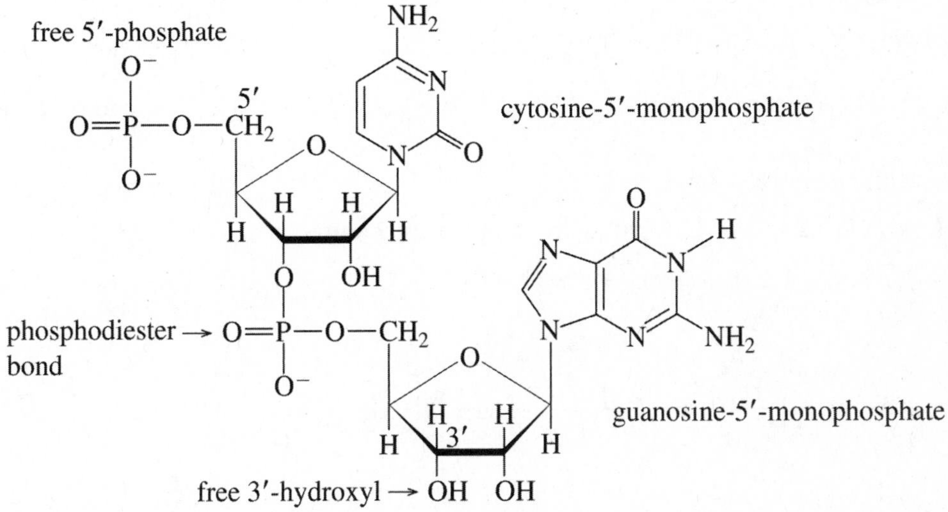

♦ **Learning Exercise 17.2B**

Consider the following sequence of nucleotides in RNA: —A—G—U—C—

1. What are the names of the nucleotides in this sequence? _____

_____

2. Which nucleotide has the free 5′-phosphate group? _____

3. Which nucleotide has the free 3′-hydroxyl group? _____

Answers

    1. adenosine-5′-monophosphate, guanosine-5′-monophosphate,
        uridine-5′-monophosphate, cytidine-5′-monophosphate

    2. adenosine-5′-monophosphate

    3. cytidine-5′-monophosphate

17.3 DNA Double Helix

- The two strands in DNA are held together by hydrogen bonds between complementary base pairs, A with T, and G with C.
- During DNA replication, DNA strands separate, and DNA polymerase makes new DNA strands along each of the original DNA strands that serve as templates.
- Complementary base pairing ensures the correct placement of bases to give identical copies of the original DNA.

(MC)

Tutorial: The Double Helix
Tutorial: DNA Replication
Self Study Activity: DNA Replication

♦ **Learning Exercise 17.3A**

Complete the following statements.

1. The structure of the two strands of nucleotides in DNA is called a _____.

2. The only base pairs that connect the two DNA strands are _____ and _____.

3. The base pairs along one DNA strand are _____ to base pairs on the opposite strand.

Answers   1. double helix   2. A—T; G—C   3. complementary

♦ **Learning Exercise 17.3B**

Complete each DNA section by writing the complementary strand:

1. —A—T—G—C—T—T—G—G—C—T—C—C—

2. —A—A—A—T—T—T—C—C—C—G—G—G—

3. —G—C—G—C—T—C—A—A—A—T—G—C—

Answers   1. —T—A—C—G—A—A—C—C—G—A—G—G—
                2. —T—T—T—A—A—A—G—G—G—C—C—C—
                3. —C—G—C—G—A—G—T—T—T—A—C—G—

♦ **Learning Exercise 17.3C**

Essay: Describe the process by which DNA replicates to produce identical copies.

Answer   In the replication process, the bases on each strand of the separated parent DNA are paired with their complementary bases. Because each complementary base is specific for a base in DNA, the new DNA strands exactly duplicate the original strands of DNA.

17.4 RNA and the Genetic Code

- The three types of RNA differ by function in the cell: ribosomal RNA makes up most of the structure of the ribosomes, messenger RNA carries genetic information from the DNA to the ribosomes, and transfer RNA places the correct amino acids in the protein.
- Transcription is the process by which RNA polymerase produces mRNA from one strand of DNA.
- The bases in the mRNA are complementary to the DNA, except U is paired with A in DNA.
- The production of mRNA occurs when certain proteins are needed in the cell.
- The genetic code consists of a sequence of three bases (triplet) that specifies the order for the amino acids in a protein.
- There are 64 codons for the 20 amino acids, which means there are several codons for most amino acids.
- The AUG codon signals the start of transcription and UAG, UGA, and UAA codons signal stop.

(MC)

Tutorial: Types of RNA
Tutorial: Transcription
Tutorial: Genetic Code
Tutorial: Following the Instructions in DNA

◆ Learning Exercise 17.4A

Match each of the following characteristics with a specific type of RNA: mRNA, tRNA, or rRNA:

1. the most abundant type of RNA in a cell _____

2. the RNA that has the shortest chain of nucleotides _____

3. the RNA that carries information from DNA to the ribosomes for protein synthesis _____

4. the RNA that is the major component of ribosomes _____

5. the RNA that carries specific amino acids to the ribosome for protein synthesis _____

6. the RNA that consists of a large and a small subunit _____

Answers   1. rRNA   2. tRNA   3. mRNA
              4. rRNA   5. tRNA   6. rRNA

◆ Learning Exercise 17.4B

Fill in the blanks with a word or phrase that answers each of the following questions:

1. Where in the cell does transcription take place?   _____

2. How many strands of the DNA molecules are involved?   _____

3. The abbreviations for the four nucleotides in mRNA are   _____

4. Write the corresponding section of mRNA produced from each of the following strands of DNA.

a. —C—A—T—T—C—G—G—T—A—

b. —G—T—A—C—C—T—A—A—C—G—T—C—C—G—

Answers   **1.** nucleus   **2.** one   **3.** A, U, G, C
                       4. a. —G—U—A—A—G—C—C—A—U—
                            b. —C—A—U—G—G—A—U—U—G—C—A—G—G—C—

◆ **Learning Exercise 17.4C**

Indicate the amino acid coded for by the following mRNA codons.

1. UUU _____   **2.** ACA _____

3. GCG _____   **4.** AUG _____

5. AGC _____   **6.** CUC _____

7. CCA _____   **8.** CAU _____

9. GGA _____   **10.** GUU _____

Answers   **1.** Phe   **2.** Thr   **3.** Ala   **4.** Start/Met **5.** Ser
                   6. Leu   **7.** Pro   **8.** His   **9.** Gly   **10.** Val

17.5 Protein Synthesis

- Proteins are synthesized at the ribosomes in a translation process that includes three steps: initiation, elongation, and termination.
- During translation, the different tRNA molecules bring the appropriate amino acids to the ribosome where the amino acid is bonded by a peptide bond to the growing peptide chain.
- When the polypeptide is released, it takes on its secondary and tertiary structures to become a functional protein in the cell.

Self Study Activity: Overview of Protein Synthesis
Tutorial: Protein Synthesis
Self Study Activity: Translation
Tutorial: Predicting an Amino Acid Sequence

◆ **Learning Exercise 17.5A**

Match the components of the translation process **a–e** with the statements **1–5**.

    a. initiation   **b.** activation   **c.** anticodon   **d.** translocation   **e.** termination

1. _____ the three bases in each tRNA that complement a codon on the mRNA

2. _____ the combining of an amino acid with a specific tRNA

3. _____ the placement of methionine on the large ribosome

4. _____ the shift of the ribosome from one codon on mRNA to the next

5. _____ the process that occurs when the ribosome reaches a UAA, UAG, or UGA codon on mRNA

Answers   **1.** c   **2.** b   **3.** a   **4.** d   **5.** e

♦ **Learning Exercise 17.5B**

Write the mRNA that would form for the following section of DNA. For each codon in the mRNA, write the amino acid that would be placed in the protein by a tRNA.

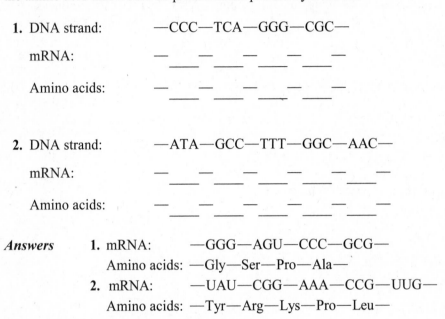

1. DNA strand: —CCC—TCA—GGG—CGC—

   mRNA: — __ — __ — __ — __ —

   Amino acids: — __ — __ — __ — __ —

2. DNA strand: —ATA—GCC—TTT—GGC—AAC—

   mRNA: — __ — __ — __ — __ — __ —

   Amino acids: — __ — __ — __ — __ — __ —

Answers   **1.** mRNA: —GGG—AGU—CCC—GCG—
               Amino acids: —Gly—Ser—Pro—Ala—
               2. mRNA: —UAU—CGG—AAA—CCG—UUG—
               Amino acids: —Tyr—Arg—Lys—Pro—Leu—

♦ **Learning Exercise 17.5C**

A segment of DNA that codes for a protein contains 270 nucleotides. How many amino acids would be present in the protein produced from this DNA segment?

Answer   Assuming that the entire segment codes for a protein, there would be $90 \, (270 \div 3)$ amino acids in the protein produced.

17.6 Genetic Mutations

- A genetic mutation is a change of one or more bases in the DNA sequence that may alter the structure and ability of the resulting protein to function properly.
- In a substitution, one base is altered, which may code for a different amino acid.
- In a frameshift mutation, the insertion or deletion of one base alters all the codons following the base change, which affects the amino acid sequence that follows the mutation.

Tutorial: Genetic Mutations
Tutorial: Mutations and Genetic Diseases

♦ **Learning Exercise 17.6**

Consider the DNA template: —AAT—CCC—GGG—

1. Write the mRNA produced.

    — — — —
    ____ ____ ____

2. Write the amino acid order for the mRNA codons.

    — — — —
    ____ ____ ____

3. A mutation replaces the thymine in the DNA template with guanine. Write the mRNA it produces.

    — — — —
    ____ ____ ____

4. Write the new amino acid order.

    — — — —
    ____ ____ ____

5. Why is this mutation called a substitution?

6. How is a substitution different from a frameshift mutation?

7. What are some possible causes of mutations?

Answers

1. —UUA—GGG—CCC— 2. —Leu—Gly—Pro—
3. —UUC—GGG—CCC— 4. —Phe—Gly—Pro—
5. In a substitution, one base is replaced with another; only one codon is affected and only one amino acid may be different.
6. In a frameshift mutation, the codons that follow the mutation are shifted by one base, which causes a different amino acid order in the remaining sequence of the protein.
7. X-rays, UV light, chemicals called mutagens, and some viruses are possible causes of mutations.

17.7 Viruses

- Viruses are small particles of 3–200 genes that cannot replicate unless they invade a host cell.
- A viral infection involves using the host cell machinery to replicate the viral DNA.
- A retrovirus contains RNA and a reverse transcriptase enzyme that synthesizes a viral DNA in a host cell.

Tutorial: HIV Reproductive Cycle

♦ **Learning Exercise 17.7**

Match the key terms with the statements shown below.

    a. host cell   **b.** retrovirus   **c.** vaccine   **d.** protease   **e.** virus

1. _____ the enzyme inhibited by drugs that prevent the synthesis of viral proteins

2. _____ a small, disease-causing particle that contains either DNA or RNA as its genetic material

3. _____ a type of virus that must use reverse transcriptase to make a viral DNA

4. _____ required by viruses to replicate

5. _____ inactive form of viruses that boosts the immune response by causing the body to produce antibodies

Answers   **1.** d   **2.** e   **3.** b   **4.** a   **5.** c

Checklist for Chapter 17

You are ready to take the Practice Test for Chapter 17. Be sure that you have accomplished the following learning goals for this chapter. If you are not sure, review the section listed at the end of the goal. Then apply your new skills and understanding to the Practice Test.

After studying Chapter 17, I can successfully:

_____ Identify the components of nucleic acids RNA and DNA (17.1).

_____ Describe the nucleotides contained in DNA and RNA (17.1).

_____ Describe the primary structure of nucleic acids (17.2).

_____ Describe the structure of RNA and DNA; show the relationship between the bases in the double helix (17.3).

_____ Explain the process of DNA replication (17.3).

_____ Describe the structures and characteristics of the three types of RNA (17.4).

_____ Describe the synthesis of mRNA (transcription) (17.4).

_____ Describe the function of the codons in the genetic code (17.4).

_____ Describe the role of translation in protein synthesis (17.5).

_____ Describe some ways in which DNA is altered to cause mutations (17.6).

_____ Explain how retroviruses use reverse transcription to synthesize DNA (17.7).

Practice Test for Chapter 17

1. A nucleotide contains
    A. a base
    B. a base and a sugar
    C. a phosphate and a sugar
    D. a base and a deoxyribose
    E. a base, a sugar, and a phosphate

2. The double helix in DNA is held together by
    A. hydrogen bonds   **B.** ester linkages   **C.** peptide bonds
    D. salt bridges   **E.** disulfide bonds

3. The process of producing DNA in the nucleus is called
   A. complementation   **B.** replication   **C.** translation
   D. transcription   **E.** mutation

4. Which occurs in RNA but *not* in DNA?
   A. thymine   **B.** cytosine   **C.** adenine   **D.** phosphate   **E.** uracil

5. Which molecule determines protein structure in protein synthesis?
   A. DNA   **B.** mRNA   **C.** tRNA   **D.** rRNA   **E.** ribosomes

6. Which type of molecule carries amino acids to the ribosomes?
   A. DNA   **B.** mRNA   **C.** tRNA   **D.** rRNA   **E.** protein

For questions 7 through 15, select answers from the following nucleic acids:
   A. DNA   **B.** mRNA   **C.** tRNA   **D.** rRNA

7. _____ along with protein, it is a major component of the ribosomes

8. _____ a double helix consisting of two chains of nucleotides held together by hydrogen bonds between bases

9. _____ a nucleic acid that uses deoxyribose as the sugar

10. _____ a nucleic acid produced in the nucleus that migrates to the ribosomes to direct the formation of a protein

11. _____ it can place the proper amino acid into the peptide chain

12. _____ it has the bases adenine, cytosine, guanine, and thymine

13. _____ it contains the codons for the amino acid order

14. _____ it contains a triplet called an anticodon loop

15. _____ this nucleic acid is replicated during cellular division

For questions 16 through 20, select answers from the following:

A. —A—G—C—C—T—A—   **B.** —A—U—U—G—C—U—C—
        | | | | | |
      —T—C—G—G—A—T—

C. —A—G—T—U—G—U—   **D.** —G—U—A—
        | | | | | |
      —T—C—A—A—C—A—

E. —A—T—G—T—A—T—

16. _____ a section of mRNA   **17.** _____ an impossible section of DNA

18. _____ a codon   **19.** _____ a section from a DNA molecule

20. _____ a single strand that would not be possible for mRNA

Use the statements A–E to answer questions 21 through 25.

    A. tRNA assembles the amino acids at the ribosomes
    B. DNA forms a complementary copy of itself called mRNA
    C. protein is formed and breaks away
    D. tRNA picks up specific amino acids
    E. mRNA goes to the ribosomes

21. _____ first step

22. _____ second step

23. _____ third step

24. _____ fourth step

25. _____ fifth step

Answers to the Practice Test

1. E	**2.** A	**3.** B	**4.** E	**5.** A
6. C	**7.** D	**8.** A	**9.** A	**10.** B
11. C	**12.** A	**13.** B	**14.** C	**15.** A
16. B, D	**17.** C	**18.** D	**19.** A	**20.** E
21. B	**22.** E	**23.** D	**24.** A	**25.** C

18

Metabolic Pathways and Energy Production

Study Goals

- Explain the role of ATP in anabolic and catabolic reactions.
- Compare the structures and function of the coenzymes NAD^+, FAD, and coenzyme A.
- Give the sites, enzymes, and products for the digestion of carbohydrates, triacylglycerols, and proteins.
- Describe the key reactions in the degradation of glucose in glycolysis.
- Describe the pathways for pyruvate.
- Describe the reactions in the citric acid cycle that oxidize acetyl CoA.
- Explain how electrons from NADH and H^+ and FAD move through electron transport to form H_2O.
- Describe the role of oxidative phosphorylation in ATP synthesis.
- Calculate the ATP produced by the complete combustion of glucose.
- Describe the oxidation of fatty acids via β oxidation.
- Calculate the ATP produced by the complete oxidation of a fatty acid.
- Explain the role of transamination and oxidative deamination in the degradation of amino acids.

Think About It

1. Why do you need ATP in your cells?

2. What monosaccharides are produced when carbohydrates undergo digestion?

3. What is meant by *aerobic* and *anaerobic* conditions in the cells?

4. Why is the citric acid cycle considered a central pathway in metabolism?

5. What are the products from the digestion of triacylglycerols and proteins?

6. When do you utilize fats and proteins for energy?

Key Terms

1. Match the key terms with the correct statement shown below.

    a. ATP   **b.** anabolic reaction   **c.** glycolysis
    d. catabolic reaction   **e.** mitochondria

1. _____ a metabolic reaction that uses energy to build large molecules

2. _____ a metabolic reaction that produces energy for the cell by degrading large molecules

3. _____ a high-energy compound that provides energy for reactions

4. _____ the degradation reactions of glucose that yield two pyruvate molecules

5. _____ the organelles in the cells where energy-producing reactions take place

Answers   **1.** b   **2.** d   **3.** a   **4.** c   **5.** e

2. Match the key terms with the correct statement shown below:

    a. citric acid cycle   **b.** oxidative phosphorylation   **c.** coenzyme Q
    d. cytochrome *c*   **e.** proton pump

1. _____ a mobile carrier that passes electrons from NADH and $FADH_2$ to complex III

2. _____ a mobile electron carrier that transfers electrons from complex III to complex IV

3. _____ protons move from the matrix into the inner membrane space to create a proton gradient

4. _____ the synthesis of ATP from ADP and P_i using energy generated from electron transport

5. _____ oxidation reactions that convert acetyl CoA to CO_2 producing reduced coenzymes for energy production via electron transport

Answers   **1.** c   **2.** d   **3.** e   **4.** b   **5.** a

18.1 Metabolism and ATP Energy

- Metabolism is all the chemical reactions that provide energy and substances for cell growth.
- Catabolic reactions degrade large molecules to produce energy.
- Anabolic reactions utilize energy in the cells to build large molecules for the cells.
- Energy is stored in ATP, a high-energy compound that is hydrolyzed when energy is required for the anabolic reactions that do work in the cells.
- The hydrolysis of ATP, which releases energy, is linked with many anabolic reactions in the cell.

(MC)

Tutorial: Metabolism and Cell Structure
Tutorial: ATP: Energy Storage

♦ **Learning Exercise 18.1A**

Identify the stages of metabolism for each of the following processes:

    a. Stage 1   **b.** Stage 2   **c.** Stage 3

1. _____ Oxidation of two-carbon acetyl CoA provides most of the energy for ATP synthesis.

2. _____ Polysaccharides undergo digestion to monosaccharides such as glucose.

3. _____ Digestion products such as glucose are degraded to two- or three-carbon compounds.

Answers   **1.** c   **2.** a   **3.** b

♦ **Learning Exercise 18.1B**

Complete the following statements for ATP:
The ATP molecule is composed of a base (1) _____, a (2) _____ sugar, and three
(3) _____. ATP undergoes (4) _____, which cleaves a (5) _____ and
releases (6) _____. For this reason, ATP is called a (7) _____ compound. The
resulting phosphate group, called inorganic phosphate, is abbreviated as (8) _____. This
equation can be written as (9) _____. The energy from ATP is linked
to cellular reactions that are (10) _____.

Answers   **1.** adenine   **2.** ribose   **3.** phosphate groups   **4.** hydrolysis
                5. phosphate   **6.** energy   **7.** high-energy   **8.** P_i
                9. $ATP + H_2O \rightarrow ADP + P_i + energy$ (7.3 kcal/mole)   **10.** energy requiring (anabolic)

18.2 Digestion of Foods

- Digestion is a series of reactions that break down large food molecules of carbohydrates, lipids, and proteins into smaller molecules that can be absorbed and used by the cells.
- The end products of digestion of polysaccharides are the monosaccharides glucose, fructose, and galactose.
- Proteins begin digestion in the stomach, where HCl denatures proteins and activates peptidases that hydrolyze peptide bonds.
- In the small intestine, trypsin and chymotrypsin complete the hydrolysis of peptides to amino acids.
- Dietary fats begin digestion in the small intestine where they are emulsified by bile salts.
- Pancreatic lipases hydrolyze the triacylglycerols to yield monoacylglycerols and free fatty acids.
- The triacylglycerols are reformed in the intestinal lining, where they combine with proteins to form chylomicrons for transport through the lymph system and bloodstream.
- In the cells, triacylglycerols are hydrolyzed to glycerol and fatty acids, which can be used for energy.

(MC)

Tutorial: Breakdown of Carbohydrates
Tutorial: Digestion of Triacylglycerols
Tutorial: Protein Digestion

♦ Learning Exercise 18.2A

Complete the table to describe sites, enzymes, and products for the digestion of carbohydrates.

Food	Digestion site(s)	Enzyme	Products
1. Amylose			
2. Amylopectin			
3. Maltose			
4. Lactose			
5. Sucrose			

Answers

Food	Digestion site(s)	Enzyme	Products
1. Amylose	**a.** mouth	**a.** salivary amylase	**a.** smaller polysaccharides (dextrins), some maltose and glucose
	b. small intestine	**b.** pancreatic amylase	**b.** maltose, glucose
2. Amylopectin	**a.** mouth	**a.** salivary amylase	**a.** smaller polysaccharides (dextrins), some maltose and glucose
	b. small intestine	**b.** pancreatic amylase	**b.** maltose, glucose
3. Maltose	small intestine	maltase	glucose and glucose
4. Lactose	small intestine	lactase	glucose and galactose
5. Sucrose	small intestine	sucrase	glucose and fructose

♦ Learning Exercise 18.2B

Match each of the following terms with the correct description:

a. stomach   **b.** small intestine

1. _____ HCl activates enzymes that hydrolyzes peptide bonds in proteins.

2. _____ Trypsin and chymotrypsin convert peptides to amino acids.

Answers   **1.** a   **2.** b

18.3 Coenzymes in Metabolic Pathways

- Coenzymes such as FAD and NAD^+ pick up hydrogen and electrons during oxidative processes.
- Coenzyme A carries acetyl (two-carbon) groups produced when glucose, fatty acids, and amino acids are degraded.

♦ Learning Exercise 18.3

Select the coenzyme that matches each of the following descriptions of coenzymes:

a. NAD^+ **b.** NADH   **c.** FAD   **d.** $FADH_2$ **e.** coenzyme A

1. _____ participates in reactions that convert a hydroxyl group to a C=O group

2. _____ contains riboflavin (vitamin B_2)

3. _____ reduced form of nicotinamide adenine dinucleotide

4. _____ contains the vitamin niacin

5. _____ oxidized form of flavin adenine dinucleotide

6. _____ contains the vitamin pantothenic acid, ADP, and an aminoethanethiol

7. _____ participates in oxidation reactions that produce a carbon–carbon double bond (C=C)

8. _____ transfers acyl groups such as the two-carbon acetyl group

9. _____ reduced form of flavin adenine dinucleotide

Answers   **1.** a   **2.** c, d   **3.** b   **4.** a, b   **5.** c
        6. e   **7.** c   **8.** e   **9.** d

18.4 Glycolysis: Oxidation of Glucose

- Glycolysis is the primary anaerobic pathway for the degradation of glucose to yield pyruvic acid.
- Glucose is converted to fructose-1,6-bisphosphate that is split into two triose phosphate molecules.
- The oxidation of the three-carbon sugar yields the reduced coenzyme 2 NADH, and 2 ATP.
- In the absence of oxygen, pyruvate is reduced to lactate and NAD^+ is regenerated for the continuation of glycolysis.
- Under aerobic conditions, pyruvate is oxidized and converted to acetyl CoA.

Tutorial: The Glycolysis Pathway
Tutorial: Energy Use and Capture by Glycolysis
Tutorial: Pathways for Pyruvate
Self Study Activity: Glycolysis

◆ **Learning Exercise 18.4A**

Match each of the following terms of glycolysis with the description:

   a. 2 NADH   **b.** anaerobic   **c.** glucose   **d.** two pyruvate

   e. energy invested   **f.** energy generated   **g.** 2 ATP   **h.** 4 ATP

1. _____ the starting material for glycolysis   **2.** _____ number of reduced coenzymes produced

3. _____ steps 1–5 of glycolysis   **4.** _____ steps 6–10 of glycolysis

5. _____ operates without oxygen   **6.** _____ end products of glycolysis

7. _____ net ATP energy produced   **8.** _____ number of ATP required

Answers   **1.** c   **2.** a   **3.** e   **4.** f   **5.** b
              6. a, d, g **7.** g   **8.** g

◆ **Learning Exercise 18.4B**

Fill in the blanks with the following terms:

    lactate NAD^+ acetyl CoA

    $NADH + H^+$ aerobic   anaerobic

When oxygen is available during glycolysis, the three-carbon pyruvate may be oxidized to form
(1) _____ + CO_2. The coenzyme (2) _____ is reduced to (3) _____.
Under (4) _____ conditions, pyruvate is reduced to (5) _____. Under (6) _____
conditions, pyruvate is oxidized to acetyl CoA.

Answers   **1.** acetyl CoA   **2.** NAD^+ **3.** $NADH + H^+$
              4. anaerobic   **5.** lactate   **6.** aerobic

◆ **Learning Exercise 18.4C**

Essay: Explain how the formation of lactate from pyruvate during anaerobic conditions allows glycolysis to continue.

Answer   Under anaerobic conditions, the oxidation of pyruvate to acetyl CoA to regenerate NAD^+
               cannot take place. However, pyruvate is reduced to lactate using $NADH + H^+$ in the
               cytoplasm and regenerating NAD^+.

18.5 The Citric Acid Cycle

- At the start of a sequence of reactions called the citric acid cycle, acetyl CoA combines with oxaloacetate to yield citrate.
- In one turn of the citric acid cycle, the oxidation of acetyl CoA yields two CO_2, one GTP, three NADH, and one $FADH_2$. The phosphorylation of ADP by GTP yields ATP.

(MC)

Tutorial: The Citric Acid Cycle
Self Study Activity: Krebs Cycle
Tutorial: Citric Acid Cycle

◆ **Learning Exercise 18.5**

In each of the following steps of the citric acid cycle, indicate if oxidation occurs (yes/no) and any coenzyme or direct phosphorylation product produced ($NADH + H^+$, $FADH_2$, GTP):

Steps in citric acid cycle	Oxidation	Coenzyme
1. acetyl CoA + oxaloacetate $\rightarrow$ citrate	_____	_____
2. citrate $\rightarrow$ isocitrate	_____	_____
3. isocitrate $\rightarrow \alpha$-ketoglutarate	_____	_____
4. α-ketoglutarate $\rightarrow$ succinyl CoA	_____	_____
5. succinyl CoA $\rightarrow$ succinate	_____	_____
6. succinate $\rightarrow$ fumarate	_____	_____
7. fumarate $\rightarrow$ malate	_____	_____
8. malate $\rightarrow$ oxaloacetate	_____	_____

Answers   1. no   2. no   3. yes, $NADH + H^+$ 4. yes, $NADH + H^+$
5. no, GTP   6. yes, $FADH_2$ 7. no   8. yes, $NADH + H^+$

18.6 Electron Transport and Oxidative Phosphorylation

- The reduced coenzymes from glycolysis and the citric acid cycle are oxidized to NAD^+ and FAD by transferring protons and electrons to electron transport.
- In electron transport, electrons are transferred to electron carriers including coenzyme Q, and cytochrome *c*.
- The final acceptor, O_2, combines with protons and electrons to yield H_2O.
- The flow of electrons through electron transport pumps protons across the inner membrane, which produces a high-energy proton gradient that provides energy for the synthesis of ATP.
- The process of using the energy of electron transport to synthesize ATP is called oxidative phosphorylation.
- The oxidation of NADH yields three ATP molecules, and $FADH_2$ yields two ATP.
- The complete oxidation of glucose yields a total of 36 ATP from direct phosphorylation and the oxidation of the reduced coenzymes NADH and $FADH_2$.

♦ **Learning Exercise 18.6A**

Write the oxidized and reduced forms of each of the following electron carriers.

1. flavin adenine dinucleotide   oxidized _____   reduced _____

2. coenzyme Q   oxidized _____   reduced _____

3. cytochrome c oxidized _____   reduced _____

Answers   1. oxidized: FAD   reduced: $FADH_2$

                    2. oxidized: Q   reduced: QH_2

                    3. oxidized: cyt c (Fe^{3+})   reduced: cyt c (Fe^{2+})

(MC)

Tutorial: Electron Transport
Tutorial: Power from Protons: ATP Synthase
Tutorial: The Chemiosmotic Model
Tutorial: ATP Energy from Glucose
Self Study Activity: Electron Transport

♦ **Learning Exercise 18.6B**

1. Write an equation for the transfer of hydrogen from $FADH_2$ to Q.

2. What is the function of coenzyme Q in electron transport?

3. What are the end products of electron transport?

Answers   1. $FADH_2 + Q \rightarrow FAD + QH_2$

                  2. Q accepts hydrogen atoms from NAD^+ or $FADH_2$. From QH_2, the hydrogen atoms are separated into protons and electrons, with the electrons being passed on to cytochrome c.

                  3. H_2O

♦ **Learning Exercise 18.6C**

Match the following terms with the correct description below:

a. oxidative phosphorylation   b. ATP synthase

c. proton pumps   d. proton gradient

1. _____ the complexes I, III, and IV through which H^+ ions move out of the matrix into the intermembrane space

2. _____ the protein channel where protons flow from the intermembrane space back to the matrix, generating energy for ATP synthesis

3. _____ energy from electron transport is used to form a proton gradient that drives ATP synthesis

4. _____ the accumulation of protons in the intermembrane space that lowers pH

Answers   **1.** c   **2.** b   **3.** a   **4.** d

♦ **Learning Exercise 18.6D**

Complete the following:

Substrate	Reaction	Products	ATP produced
1. Glucose	glycolysis (aerobic)		
2. Pyruvate	oxidation		
3. Acetyl CoA	citric acid cycle		
4. Glucose	glycolysis (anaerobic)		
5. Glucose	complete oxidation		

Answers

Substrate	Reaction	Products	ATP produced
1. Glucose	glycolysis (aerobic)	2 pyruvate $+ 2H_2O$	6 ATP
2. Pyruvate	oxidation	acetyl CoA $+ CO_2$	3 ATP
3. Acetyl CoA	citric acid cycle	$2CO_2 + 2H_2O$	12 ATP
4. Glucose	glycolysis (anaerobic)	2 lactate	2 ATP
5. Glucose	complete oxidation	$6CO_2 + 6H_2O$	36 ATP

18.7 Oxidation of Fatty Acids

- When needed for energy, fatty acids combine with coenzyme A for transport to the mitochondria where they undergo β oxidation.
- In β oxidation, a fatty acyl chain is oxidized to yield a shortened fatty acid, acetyl CoA, and the reduced coenzymes NADH and $FADH_2$.
- The energy obtained from a particular fatty acid depends on the number of carbon atoms.
- Two ATP are required for activation. Then each acetyl CoA produces 12 ATP via the citric acid cycle, and electron transport converts each NADH to 3 ATP and each FADH to 2 ATP.

Tutorial: Oxidation of Fatty Acids
Tutorial: Energy from Fatty Acid Oxidation
Tutorial: Ketogenesis and Ketone Bodies
Case Study: Diabetes and Blood Glucose

♦ **Learning Exercise 18.7A**

1. Write an equation for the activation of myristic (C_{14}) acid: $CH_3-(CH_2)_{12}-\overset{\overset{\textstyle O}{\|}}{C}-OH$.

2. Write an equation for the first oxidation of myristyl CoA.

3. Write an equation for the hydration of the double bond.

4. Write the overall equation for the complete oxidation of myristyl CoA.

5. **a.** How many cycles of β oxidation will be required?

   b. How many acetyl CoA units will be produced?

   c. What is the total ATP produced by complete oxidation?

1. $CH_3-(CH_2)_{12}-\overset{\overset{\displaystyle O}{\|}}{C}-OH + HS-CoA + ATP \rightarrow CH_3-(CH_2)_{12}-\overset{\overset{\displaystyle O}{\|}}{C}-S-CoA + AMP + 2P_i + H_2O$

2. $CH_3-(CH_2)_{12}-\overset{\overset{\displaystyle O}{\|}}{C}-S-CoA + FAD \rightarrow CH_3-(CH_2)_{10}-CH=CH-\overset{\overset{\displaystyle O}{\|}}{C}-S-CoA + FADH_2$

3. $CH_3-(CH_2)_{10}-CH=CH-\overset{\overset{\displaystyle O}{\|}}{C}-S-CoA + H_2O \rightarrow CH_3-(CH_2)_{10}-\overset{\overset{\displaystyle OH}{|}}{CH}-CH_2-\overset{\overset{\displaystyle O}{\|}}{C}-S-CoA$

4. myristyl $(C_{14})-CoA + 6\,CoA + 6\,FAD + 6\,NAD^+ + 6\,H_2O \rightarrow$

$$7\,Acetyl\,CoA + 6\,FADH_2 + 6\,NADH + 6\,H^+$$

5. **a.** 6 cycles   **b.** 7 acetyl CoA   **c.** 112 ATP

♦ **Learning Exercise 18.7B**

Lauric acid is a 12-carbon saturated fatty acid: $CH_3-(CH_2)_{10}-COOH$.

1. How many ATP are needed for activation?

2. How many cycles of β oxidation are required?

3. How many NADH and $FADH_2$ are produced during β oxidation?

4. How many acetyl CoA units are produced?

5. What is the total ATP produced from electron transport and the citric acid cycle?

Answers   **1.** 2 ATP   **2.** 5 cycles
               3. 5 cycles produce 5 NADH and 5 $FADH_2$ **4.** 6 acetyl CoA
               5. 5 NADH×3 ATP=15 ATP; 5 $FADH_2$×2 ATP=10 ATP;
                     6 acetyl CoA×12 ATP = 72 ATP; Total ATP =15 ATP+10 ATP+72 ATP −
                     2 ATP (for activation) = 95 ATP

18.8 Degradation of Amino Acids

- Amino acids are normally used for protein synthesis.
- When needed for energy, amino acids are degraded by transferring an amino group from an amino acid to an α-keto acid to yield a different amino acid and α-keto acid.
- In oxidative deamination, the amino group in glutamate is removed as an ammonium ion, NH_4^+.
- α-Keto acids resulting from transamination can be used as intermediates in the citric acid cycle, in the synthesis of lipids or glucose, or oxidized for energy.

♦ Learning Exercise 18.8A

Match each of the following descriptions with transamination (T) or oxidative deamination (D):

1. _____ produces an ammonium ion, NH_4^+

2. _____ transfers an amino group to an α-keto acid

3. _____ usually involves the degradation of glutamate

4. _____ requires NAD^+

5. _____ produces another amino acid and α-keto acid

6. _____ usually produces α-ketoglutarate

Answers　　　1. D　　　2. T　　　3. D　　　4. D　　　5. T　　　6. D

Tutorial: **Transamination and Deamination**
Tutorial: **Detoxifying Ammonia in the Body**
Tutorial: **The Fate of Carbon Atoms in Amino Acids**

♦ Learning Exercise 18.8B

1. Write an equation for the transamination reaction of serine and oxaloacetate.

2. Write an equation for the oxidative deamination of glutamate.

1. $\underset{\underset{\displaystyle HO-CH_2-CH-COO^-}{|}}{\overset{\overset{\displaystyle NH_3{}^+}{|}}{}} + \underset{\displaystyle {}^-OOC-\overset{\overset{\displaystyle O}{\|}}{C}-CH_2-COO^-}{} \xrightarrow{\quad\text{Aminotransferase}\quad}$

$\qquad\qquad HO-CH_2-\overset{\overset{\displaystyle O}{\|}}{C}-COO^- + {}^-OOC-\underset{\underset{\displaystyle NH_3{}^+}{|}}{CH}-CH_2-COO^-$

2. $\underset{\underset{\displaystyle NH_3{}^+}{|}}{{}^-OOC-CH}-CH_2-CH_2-COO^- + NAD^+ + H_2O \xrightarrow{\quad\substack{\text{Glutamate}\\ \text{dehydrogenase}}\quad}$

$\qquad\qquad {}^-OOC-\overset{\overset{\displaystyle O}{\|}}{C}-CH_2-CH_2-COO^- + NH_4{}^+ + NADH + H^+$

Checklist for Chapter 18

You are ready to take the Practice Test for Chapter 18. Be sure that you have accomplished the following learning goals for this chapter. If you are not sure, review the section listed at the end of the goal. Then apply your new skills and understanding to the Practice Test.

After studying Chapter 18, I can successfully:

_____ Describe the role of ATP in catabolic and anabolic reactions (18.1).

_____ Describe the sites, enzymes, and products of digestion for carbohydrates (18.2).

_____ Describe the coenzymes NAD^+, FAD, and coenzyme A (18.3).

_____ Describe the conversion of glucose to pyruvate in glycolysis (18.4).

_____ Give the conditions for the conversion of pyruvate to lactate and acetyl coenzyme A (18.4).

_____ Describe the oxidation of acetyl CoA in the citric acid cycle (18.5).

_____ Identify the electron carriers in electron transport (18.6).

_____ Describe the process of electron transport (18.6).

_____ Explain the chemiosmotic theory whereby ATP synthesis is linked to the energy of electron transport and a proton gradient (18.6).

_____ Account for the ATP produced by the complete oxidation of glucose (18.6).

_____ Describe the oxidation of fatty acids via β oxidation (18.7).

_____ Calculate the ATP produced by the complete oxidation of a fatty acid (18.7).

_____ Explain the role of transamination and oxidative deamination in degrading amino acids (18.8).

Practice Test for Chapter 18

1. The main function of the mitochondria is
   A. energy production   B. protein synthesis   C. glycolysis
   D. genetic instructions   E. waste disposal

2. ATP is a(n)
   A. nucleotide unit in RNA and DNA
   B. end product of glycogenolysis
   C. end product of transamination
   D. enzyme
   E. energy storage molecule

For questions 3 through 6, identify one or more of the following enzymes and end products for the digestion of each:
   A. maltase   B. glucose   C. fructose   D. sucrase
   E. galactose   F. lactase   G. pancreatic amylase

3. sucrose _____   4. lactose _____

5. small polysaccharides _____   6. maltose _____

7. The digestion of triacylglycerols takes place in the _____ by enzymes called
   _____.
   A. small intestine; peptidases   B. stomach; lipases   C. stomach; peptidases
   D. small intestine; lipases   E. all of these

8. The products of the digestion of triacylglycerols are
   A. fatty acids   B. monoacylglycerols   C. glycerol
   D. diacylglycerols   E. all of these

9. The digestion of proteins takes place in the _____ by enzymes called _____.
   A. small intestine; peptidases   B. stomach; lipases   C. stomach; peptidases
   D. small intestine; lipases   E. stomach and small intestine; proteases and peptidases

10. Glycolysis
    A. requires oxygen for the catabolism of glucose
    B. represents the aerobic sequence for glucose anabolism and ATP production
    C. represents the splitting off of glucose residues from glycogen
    D. represents the anaerobic catabolism of glucose to pyruvate
    E. produces acetyl units and ATP as end products

For questions 11 through 15, associate the following coenzyme A–E with the correct description:
   A. NAD^+ B. NADH   C. FAD   D. $FADH_2$ E. coenzyme A

11. _____ converts a hydroxyl group to a $C{=}O$ group

12. _____ reduced form of nicotinamide adenine dinucleotide

13. _____ participates in oxidation reactions that produce a carbon–carbon double bond $(C{=}C)$

14. _____ transfers acyl groups such as the two-carbon acetyl group

15. _____ reduced form of flavin adenine dinucleotide

For questions 16 through 18, answer for glycolysis:

16. _____ Number of ATP invested to oxidize one glucose molecule

17. _____ Number of ATP (net) produced from one glucose molecule

18. _____ Number of NADH produced from one glucose molecule

19. Which is true of the citric acid cycle?
    A. Acetyl CoA is converted to CO_2 and H_2O.
    B. Oxaloacetate combines with acetyl units to form citric acid.
    C. The coenzymes are NAD^+, FAD, and CoA.
    D. ATP is produced by direct phosphorylation.
    E. all of the above

For questions 20 through 24, match to the compounds A–E:

    A. malate   B. fumarate   C. succinate
    D. citrate   E. oxaloacetate

20. _____ formed when oxaloacetate combines with acetyl CoA

21. _____ H_2O adds to its double bond to form malate

22. _____ FAD removes hydrogen from this compound to form a double bond

23. _____ formed when the hydroxyl group in malate is oxidized

24. _____ the compound that is regenerated in the citric acid cycle

25. One turn of the citric acid cycle produces
    A. 3 NADH   B. 3 NADH, 1 $FADH_2$ C. 3 $FADH_2$, 1 NADH, 1 ATP
    D. 3 NADH, 1 $FADH_2$, 1 ATP, 2 CO_2 E. 1 NADH, 1 $FADH_2$, 1 ATP

26. How many electron transfers in electron transport provide sufficient energy for ATP synthesis?
    A. none   B. 1   C. 2   D. 3   E. 4

27. Electron transport
    A. produces most of the body's ATP
    B. carries oxygen to the cells
    C. produces $CO_2 + H_2O$
    D. is only involved in the citric acid cycle
    E. operates during fermentation

For questions 28 through 30, match the components of electron transport with the following activities:

    A. $NADH^+$ B. $FADH_2$ C. Q   D. cytochrome c

28. _____ a mobile carrier that transfers electrons to cytochrome c

29. _____ the coenzyme that transfers hydrogen to complex II

30. _____ electron acceptor containing iron

For questions 31 through 35, select the correct answer A–F for the number of ATP produced:

A. 2 ATP	**B.** 3 ATP	**C.** 6 ATP	**D.** 12 ATP	**E.** 24 ATP
F. 36 ATP				

31. One turn of the citric acid cycle (acetyl CoA $\rightarrow 2CO_2$)

32. Complete reaction of glucose (glucose $+ 6O_2 \rightarrow 6H_2O + 6CO_2$)

33. Produced when NADH enters electron transport

34. Produced from glycolysis (glucose $+ O_2 \rightarrow 2$ pyruvate $+ 2H_2O$)

35. Produced from oxidation of 2 pyruvate (2 pyruvate $\rightarrow 2$ acetyl CoA $+ 2CO_2$)

36. Chylomicrons formed in the intestinal lining
   A. are lipoproteins
   B. are triacylglycerols coated with proteins
   C. transport fats into the lymph system and bloodstream
   D. carry triacylglycerols to the cells of the heart, muscle, and adipose tissues
   E. all of these

For questions 37 through 40, consider β oxidation of palmitic (C_{16}) acid:

37. The number of β oxidation cycles required for palmitic (C_{16}) acid is
   A. 16   **B.** 9   **C.** 8
   D. 7   **E.** 6

38. The number of acetyl CoA groups produced by the β oxidation of palmitic (C_{16}) acid is
   A. 16   **B.** 9   **C.** 8
   D. 7   **E.** 6

39. The number of NADH and $FADH_2$ produced by the β oxidation of palmitic (C_{16}) acid is
   A. 16   **B.** 9   **C.** 8
   D. 7   **E.** 6

40. The total ATP produced by the β oxidation of palmitic (C_{16}) acid is
   A. 96   **B.** 129   **C.** 131
   D. 134   **E.** 136

41. The process of transamination
   A. is part of the citric acid cycle   **B.** converts α-amino acids to β-keto acids
   C. produces new amino acids   **D.** is not used in the metabolism of amino acids
   E. is part of the β oxidation of fats

42. The oxidative deamination of glutamate produces
   A. a new amino acid   **B.** a new β-keto acid

   C. ammonia, NH_3 **D.** ammonium ion, NH_4^+

   E. urea

43. The purpose of the urea cycle in the liver is to

   A. synthesize urea

   B. convert urea to ammonium ion, NH_4^+

   C. convert ammonium ion NH_4^+ to urea

   D. synthesize new amino acids

   E. take part in the β oxidation of fats

44. The carbon atoms from various amino acids can be used in several ways such as

   A. intermediates of the citric acid cycle **B.** formation of pyruvate

   C. synthesis of glucose   **D.** formation of ketone bodies

   E. all of these

45. Essential amino acids

   A. are not synthesized by humans   **B.** are required in the diet

   C. are excreted if in excess   **D.** include leucine, lysine, and valine

   E. all of these

Answers to the Practice Test

1. A	**2.** E	**3.** D, B, C	**4.** F, B, E	**5.** G, B
6. A, B	**7.** D	**8.** E	**9.** E	**10.** D
11. A	**12.** B	**13.** C	**14.** E	**15.** D
16. 2	**17.** 2	**18.** 2	**19.** E	**20.** D
21. B	**22.** C	**23.** E	**24.** E	**25.** D
26. D	**27.** A	**28.** C	**29.** B	**30.** D
31. D	**32.** F	**33.** B	**34.** A	**35.** C
36. E	**37.** D	**38.** C	**39.** D	**40.** B
41. C	**42.** D	**43.** C	**44.** E	**45.** E